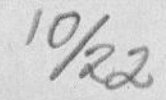

On the Trail of the Jackalope

ALSO BY MICHAEL P. BRANCH

How to Cuss in Western: And Other Missives from the High Desert

Rants from the Hill: On Packrats, Bobcats, Wildfires, Curmudgeons, a Drunken Mary Kay Lady, and Other Encounters with the Wild in the High Desert

Raising Wild: Dispatches from a Home in the Wilderness

ON THE TRAIL of the Jackalope

How a Legend Captured the World's Imagination and Helped Us Cure Cancer

MICHAEL P. BRANCH

PEGASUS BOOKS
NEW YORK LONDON

ON THE TRAIL OF THE JACKALOPE

Pegasus Books, Ltd.
148 West 37th Street, 13th Floor
New York, NY 10018

First Pegasus Books cloth edition March 2022

Interior design by Maria Fernandez

Library of Congress Cataloging-in-Publication Data is available.

ISBN: 978-1-64313-933-3

10 9 8 7 6 5 4 3 2 1

Printed in the United States of America
Distributed by Simon & Schuster
www.pegasusbooks.com

For Eryn

again and always

CONTENTS

I don't see what's so impossible about chasing two hares at once.
—Anton Chekhov

"Hallo, Rabbit," he said, "is that you?"
"Let's pretend it isn't," said Rabbit, "and see what happens."
—A. A. Milne

Ideas are like rabbits. You get a couple and learn how to handle them, and pretty soon you have a dozen.
—John Steinbeck

The truth is a rabbit in a bramble patch. And you can't lay your hand on it. All you do is circle around and point, and say, "It's in there somewhere."
—Pete Seeger

People's dreams are made out of what they do all day. The same way a dog that runs after rabbits will dream of rabbits. It's what you do that makes your soul, not the other way around.
—Barbara Kingsolver

Author's Note
Down the Rabbit Hole

I wish I could say exactly where I was when I first saw a jackalope. I was just a kid, and my initial response to the odd bunny was grinning fascination. I recall wondering if the animal was real—and hoping like hell that it was—while also realizing that, even if it wasn't, something wonderful was before my eyes. What charmed me most was the way the hybrid horned rabbit crossed boundaries, refusing to be either this thing or that thing—a form of resistance any kid struggling to navigate the adult world can appreciate. It also seemed to satirize the genre of the hunting trophy mount, a form of nature commemoration that, even as a boy, I had difficulty comprehending. The jackalope struck me as inherently playful, at once cute and funny, but still out to fool you if it could. Like all good humorists, the jackalope always keeps a straight face, taking itself seriously no matter how much it might make us laugh. The antlered bunny is a comic, to be sure, but he is pure deadpan.

Since my boyhood introduction to the jackalope, my appreciation for this bizarre creature has deepened immeasurably. The jackalope is now ubiquitous in American culture. Everywhere I travel I see jackalope mounts, jackalope kitsch and art, bumper stickers and postcards, beer and whiskey, bands and songs, teams and clubs, bars and restaurants. Unlike many other widely disseminated cultural phenomena—think Disney here—nobody owns the jackalope, and no corporation or person is entitled to control its production, distribution, consumption, or interpretation. The jackalope is like a plant whose burr catches on your sock and hitches a ride to its next home, only the plant is comic folk art and its burr instead catches your imagination, convincing you to blow ten bucks on a jackalope shot glass that you can't help but bring home. Anyone who has bought a "Save the Jackalopes" T-shirt, stuck a stamp on a tacky jackalope postcard, or shared a funny jackalope image on social media has unwittingly been a vector of the horned rabbit's viral transmission. The jackalope doesn't have a marketing department but it has no trouble getting around. Because the jackalope also migrates through narrative, I meet plenty of people who claim to have seen one and will regale me with extravagant stories that are always worth the price of a pint. As a storyteller myself, what I love about the jackalope is that there is such a rich story behind it, and yet more layers of narrative behind even that story still to be discovered.

Once endemic to the American West, the jackalope has spread far beyond its home range and now inhabits the broader culture. Embodying animal hybridity in a fascinating, comical way that tests credulity, generates legends, and captivates the imagination, the irresistible horned rabbit is a beloved staple of popular culture, folklore, and humor around the globe. But the jackalope is much more than an article of iconic kitsch. Its connection to horned hares in nature leads us beyond hoax, humor, and folk narrative into a scientific quest to save human lives by understanding the viruses that cause growths on rabbits and cancers in people.

I am seeking the real story behind the strangest, funniest, most weirdly appealing animal ever invented. I have been obsessed with this little beast for two decades, and I have at last committed myself unconditionally to its discovery. As a result, I am about to go down a fascinating rabbit hole. I am on the trail of the true tale of the jackalope. My quest is to understand how a peculiar horned rabbit born of the inventiveness of a couple of kids in Depression-era rural Wyoming ended up capturing the world's imagination—and how the study of its real-life counterpart, the horned hare, resulted in Nobel prize–winning research that ultimately led to development of the world's safest and most effective anti-cancer vaccine to date.

When tracking the jackalope, it is helpful to have an open mind and a fertile imagination. I am reminded of an exchange between Lewis Carroll's Alice and the White Queen. "One *can't* believe impossible things," insists Alice. "I daresay you haven't had much practice," replies the Queen. "Why, sometimes I've believed as many as six impossible things before breakfast." Pursuing the elusive horned rabbit will lead us from the liminal world of the trickster, back to the solid reality of nature, through the fragility of our own bodies, then home again to the inexhaustible universe of the imagination. That odd little antlered bunny you've chuckled at in a bar or gift shop in fact has a complex, fascinating, and surprising story—one that deserves to be told.

Prologue
Nature's Jackalopes

By 1932, Dr. Richard E. Shope was already a prominent microbiologist working in Princeton, New Jersey, at his own laboratory at the Rockefeller Institute, which was then among the most prestigious medical research facilities in the US. There he had conducted pioneering research on viruses, including the one responsible for an estimated 50–100 million deaths during the devastating global flu pandemic of 1918. Shope was alerted to the existence of odd growths on wild cottontails by a hunter from Cherokee, Iowa, a few hundred miles northwest of Shope's childhood home in Des Moines. Hunters who had seen the strange rabbits' unusual and sometimes grotesque "horns" correctly assumed that the animals suffered from an unknown disease. These peculiar cottontails, Shope wrote, "were said to have horn-like protuberances on the skin over various parts of their bodies. The animals were referred to popularly as 'horned' or 'warty' rabbits."

Intrigued by stories of these mysterious rabbits, Shope asked his contacts in Iowa and Kansas to procure specimens of the anomalous bunnies if they could. Soon he was being sent unusual packages from the Midwest. Shope initially received only the "warts" of a diseased rabbit, which had been mailed to him from Iowa in a glycerol solution. Soon after, however, a shipment of a dozen rabbits arrived in Princeton from Kansas. Upon investigation, Shope discovered that many of the rabbits were stricken with the same disease that had produced the warts he had already examined. He quickly expanded his study to include a total of seventy-five wild cottontails, eleven of which he found to be stricken with the disorder. Those eleven rabbits with their weird, inexplicable horns became the foundation of Shope's most important research. He had a hunch that their growths might be tumors caused by an unidentified virus, but he needed to find a way to test his unorthodox theory. Fascinated by the medical mystery before him, Shope devoted himself to exploring what afflicted the strange rabbits.

Shope used a scalpel to remove the diseased rabbits' horns, which he described as "black or grayish-black in color, well keratinized," while a cross-section of the horns revealed their "white or pinkish-white fleshy center." He then minced those growths with sterile scissors and pulverized them into a fine paste in a mortar, after which he mixed the resulting material in a saline solution to place it into suspension. Next he ran the solution through a centrifuge. After decanting the supernatant fluid from the centrifuge, he strained it through a porcelain filter so fine as to allow only viruses to pass through. Consequently, the strained fluid Shope had produced from the pulverized rabbits' horns was certain to be free of bacteria.

Next, Shope prepared his test subjects for inoculation. He shaved the sides and abdomens of healthy cottontail rabbits—both wild and domestic—and lightly scarified their skin with sterile sandpaper, causing a superficial abrasion. Finally, he used a syringe to put a few drops of the fluid he had produced onto the mild scrape, gently rubbing it in with the handle of his scalpel. Once the solution was dry, the healthy rabbits were returned to their cages, and the waiting began.

If no virus was present in the solution Shope had produced from the diseased rabbits' horns, the inoculation would have no effect on the healthy rabbits' skin. He waited a few days, and then a few days longer. Even if the cause of the strange disease was a virus, as Shope hypothesized, he had no way of knowing what its incubation period might be. On the sixth day, however, the inoculated rabbits began to show subtle signs of infection. "The first detectable lesions," wrote Shope in his groundbreaking 1933 article in *The Journal of Experimental Medicine*, "consisted in minute, barely visible elevations along the lines of scarification." After a few more days, all of the experimentally inoculated animals showed indications of disease, and within three weeks, Shope reported, the rabbits had "acquired a definitely warty appearance." Over time, those rabbits' budding growths continued to emerge, sometimes developing into tall, black horns.

The dispassionate language of Richard Shope's scientific writing obscures the excitement he must have felt in that moment, working alone in his Princeton laboratory, when he first witnessed the "barely visible elevations" that would become the once-healthy rabbits' horns. Thanks to Shope's groundbreaking work, the horn of the rabbit became the first virus-induced tumor ever to be identified in a mammal, opening the way for vitally important new research into how deadly viruses attack the body. It was a watershed moment, one which would ultimately save millions of human lives.

Chapter 1
As Real as You Want Them to Be

"Monsters" can help us by giving a tangible form to our secret fears. It is less widely appreciated today that "wonders" such as the unicorn legitimize our hopes. But all imaginary animals, and to some degree all animals, are ultimately both monsters and wonders, which assist us by deflecting and absorbing our uncertainties. It is hard to tell "imaginary animals" from symbolic, exemplary, heraldic, stylized, poetic, literary, or stereotypical ones. What is reality? Until we answer that question with confidence, a sharp differentiation between real animals and imaginary ones will remain elusive. There is some yeti in every ape, and a bit of Pegasus in every horse.

—Boria Sax, *Imaginary Animals: The Monstrous, the Wondrous, and the Human*

The town of Douglas, seat of Converse County in eastern Wyoming, is a mere dot in the expansive, sparsely populated North American interior. Perched on the western edge of the windswept Great Plains, Douglas sits quietly on the gravelly banks of the North Platte River, which was a primary route for the wagon trains that crawled toward Oregon and

California during the 1840s. Originally a frontier trading post and later a Pony Express station, the town incorporated in 1887 (with 805 residents) in response to the Fremont, Elkhorn, and Missouri Valley Railroad (FE&MV) finally reaching the remote outpost the year before. Like many rowdy frontier townships, Douglas made a good business of catering to rural laborers, and by the end of the nineteenth century, this town of only a few thousand residents reportedly had twenty-five bars. In the decades that followed, Douglas experienced the kind of boom-and-bust economic cycle that it still suffers from today. In the late 1880s, overgrazing combined with brutal winters to trigger the collapse of the beef industry. During the twentieth century, similar spikes and crashes would be fueled by commodities other than cattle: coal, oil, gas, and uranium.

Douglas Herrick was born in Douglas, Wyoming, in 1920. His little brother, Ralph, came along two years later. The Herrick family lived on a modest homestead outside of town where the boys grew up hunting the rolling prairie hills, fishing the holes in their sparkling home river, and roaming open country that extended to the horizon. Doug and Ralph Herrick hunted and fished for their own entertainment and to augment the family's meals, as did many Americans struggling through the Depression. The boys also shared an interest in taxidermy, which they studied together through a mail-order correspondence course. By the 1930s, taxidermy, which experienced a revival as naturalists as well as hunters perfected the practice, was already woven into the fabric of American life. The Boy Scouts had introduced the Taxidermy merit badge in 1911 and "stuffing animals" was soon recognized as a bona fide hobby. If the Herrick boys weren't using the post-mail taxidermy class offered by the Northwestern School of Taxidermy that operated out of Omaha, Nebraska—as I suspect they were—they would have used one very like it. In the Northwestern course, nine separate booklets, which were mailed to the amateur taxidermist in periodic installments, contained forty illustrated lessons in how to gut and tan, fabricate forms, stretch skin, preserve fur, and make realistic eyes from glass.

The jackalope's origin story—often retold and yet stubbornly unverifiable—has now entered into American folklore. One day young Doug and Ralph Herrick went out roaming the green hills hunting for small game to supplement the family supper. Having bagged a jackrabbit, the brothers returned home and tossed the hare's body onto the floor of their shop in preparation to skin it. Because they had recently butchered a small deer in the shop, a modest pair of antlers already rested on the floor. By a sheer coincidence that would change the boys' lives forever, the dead rabbit happened to slide up against the deer's horns so as to make it appear the jackrabbit sported the rack. There must then have been a long pause, during which the boys stared at the accidental amalgam, wondering what to make of it. Then big brother Doug, in a moment of inspiration, exclaimed, "Let's mount that thing!"

According to Ralph, that was in 1932—though other sources claim 1934, 1936, 1938, 1939, even 1940. If Ralph's memory was correct, he and his brother Doug would have been only about ten and twelve years old, respectively, when they invented the horned rabbit mount. It must have seemed a modest accomplishment, especially for boys who had perhaps not yet received certificates for completing their mail-order taxidermy lessons. When the brothers sold that now-legendary first antlered bunny for the princely sum of ten dollars to Roy Ball, who displayed it on the wall of the bar in his Hotel LaBonte (pronounced "La-bon-tee") in nearby Douglas, the newly created hybrid animal must have seemed to them miraculous. Using only their imagination, sense of humor, and rudimentary skills as amateur taxidermists, the Herrick boys had created something new: a hybrid animal that would go on to become the most famous, beloved, and profitable taxidermy hoax in the world. That humble Herrick homestead, out there on the limitless, rolling Wyoming prairie, was the birthplace of the jackalope. Doug Herrick died of bone and lung cancer in 2003 at age eighty-two. Ralph would pass away at age ninety on the first day of 2013. Today, so long after that moment of inception, another generation of Herricks is making jackalopes in Wyoming.

By my rough estimation there are at least one million jackalope mounts in existence, many of which keep watch over local bars, tourist traps, junk shops, greasy spoon diners, and dimly lit pool halls. Once rare, the jackalope migrated from Wyoming throughout the West and then across the nation. Antlered bunnies now adorn the walls of watering holes from Los Angeles to Seattle, Dallas to New York. And while the horned rabbit is unalloyed Americana—a genuine artifact of this country's folk culture—the mythical beast has also made its way across the oceans and around the world. What's more, the iconic jackalope mount is just the tip of the iceberg of kitsch the Herrick brothers' invention has inspired. This hoax bunny has spawned not only an endless body of comic lore, but also a thriving cottage industry worth millions—one tacky T-shirt, key chain, and postcard at a time. The Herricks' hoax has long since outgrown the gift shop and is now widely celebrated in storytelling, literature, folklore, visual art, music, film, TV, video games—and plenty more, as I was reminded the other day when I spotted a whimsical jackalope tattoo gracing the shoulder of a young woman hoisting a pint of hazy IPA at my local brewpub.

The jackalope is an oddball even among oddballs. Despite the animal's legendary moniker—a portmanteau of *jackrabbit* and *antelope*—the jackalope is often made from the head of a cottontail rather than a jackrabbit and mounts are rarely fabricated using the horns of a pronghorn antelope, the wider availability of deer antlers making them the preferred choice. To complicate matters, the jackrabbit (genus *Lepus*) is not a rabbit but rather a hare, while the pronghorn (*Antilocapra americana*) is not an antelope but instead an artiodactyl ungulate indigenous to North America—the sole survivor of a dozen fantastic Antilocaprid species which, before the sweeping wave of Pleistocene extinctions broke over them, roamed the territory now inhabited by jackalopes. But when you're inventing an animal, the sound and feel of its name counts for more than taxonomic precision. "Cottonamuledeer" just doesn't have much of a ring to it.

I have just driven Alkali, my old pickup truck, fifteen hours and a thousand miles from my home in the Great Basin Desert of northwestern Nevada to Douglas, out on the windswept western edge of the Great Plains. I slowly navigate the town's wide, quiet streets until I roll up in front of the Hotel LaBonte, a lovely old brick building graced with a faded opulence. The hotel's name commemorates the LaBonte Pony Express Station on the Oregon Trail, which was located near Douglas and named after a local hunter. When this historic inn opened in 1914, its claim to fame was that every room had electricity. Now it looks like the set of a quirky Wes Anderson film. I make my way through a narrow hallway and into the hotel lobby, where a massive bison head looms over me, its glossy, black eyes seeming to follow my steps as I cross a large turquoise Navajo rug. The surrounding walls are adorned with cattle and bison art of assorted vintages, shapes, and colors, though no two pieces seem related to each other. Above the mounted head of a longhorn steer hangs an American flag. On the starburst-patterned parquet of an old mahogany table, the local newspaper is open, as if a resident ghost has left it there while fetching a tin cup of cowboy coffee from the bunkhouse.

I make my way to the beautifully carved main desk, complete with frosted glass you might see fronting a bank teller or apothecary in an old Frank Capra film. I find its well-worn glory appealing. There is apparently no one working the desk tonight, and no bell or buzzer either—just an empty wooden office chair and, above it, where an oil painting of the old hotel's esteemed patriarch might be expected to hang, a delightfully amateurish painting in an ornate, heavily gilded frame. From the neck down, the painting seems to depict a reputable schoolmarm, with her candy-striped white blouse, prim neck bow, and neatly knitted maroon vest. But emerging from the fancy collar is the cute furry face of a rabbit, complete with unblinking dark eyes and whiskers that almost twitch. Huge cupped jackrabbit ears tower above the bunny's rounded face. Protruding from its forehead is a fine pair of branching antlers.

Wandering back outside I notice that adjacent to this abandoned lobby is the bar, which, as I enter, is doing pretty well for a weeknight. A postseason baseball game plays on the television. A half-dozen folks occupy a short row of barstools, while the resounding crack of billiard balls announces that others are shooting pool in an adjoining room. Given that Douglas is a town of only six thousand, I'm surprised not to receive the stare down that usually follows the stranger in town. But folks here are cheerfully preoccupied with the ballgame and each other's company. Some, wearing Wrangler jeans and snap-button shirts, look like working ranch hands, an impression confirmed by several well-worn Stetsons hanging on a nearby hat rack. Others, in dung-brown Carhartt work clothes and mesh-backed tractor caps, are more likely roughnecks in from the oil patch to catch a cold one and a few innings between shifts. All of them, men and women alike, look like hunters.

I greet the barkeep, a young woman who draws me a pint of Bunny Hop IPA and serves it with some friendly conversation.

"You in from the patch?" she asks.

"Nope. Just passing through. I've always wanted to see this place."

She stares at me a moment, trying to reckon my angle—and then stares harder when I pull out my leather-bound journal and begin making notes. "Well, that patch is blowing up right now," she says, cautiously. "People coming in here from all over. Powder River Basin is the next big thing. Gonna be thousands more oil and gas wells out there, going in all around the old coal pits and uranium mines."

I ask her how the townspeople feel about the new energy development boom. "We need the jobs, but it'll change this place. We used to be a cattle town. And the old coal and uranium operations left us in the dust when they played out. But we need those jobs. Without the patch, kids just leave this place when they're of age, and there won't be much left here but grandparents." She pauses. "And the state fair."

I nod and change the subject before I'm outed as an environmentalist. On my first night in town I'm not looking to pick a fight with people who are just trying to get by, however much I deplore the damage the latest get-it-and-get-out scheme is doing to this beautiful prairieland.

"I tried to check in earlier, but nobody was around," I say. Now she smiles.

"I guess you're Mike? Bill said you'd be along, but you're the only one coming in tonight so he went home to catch the Dodgers-Brewers game. I got your key when you want it. So, you're the jackalope guy, right? Long way to come from Nevada for jackalopes."

"Yup, that's me. I had to see this place for myself," I tell her. "The LaBonte is kind of a jackalope holy site. The first mount ever made was displayed above this bar. A guy named Roy Ball, who owned the place back in the day, bought that first jackalope for ten bucks from Doug and Ralph Herrick during the Great Depression. The story goes that it was hanging up in here for a long time, until it vanished during the late seventies and never turned up again."

"Bunch of Herricks still in town. And, yeah, I've heard that story, too, but I don't know for sure," she says. "I think the bar maybe used to be in a different part of the building. A lot of stuff got moved around after there was a fire in the hotel." We both scan the walls, trying to imagine the spot from which that first antlered bunny presided over the bar.

"Anyhow, we love our jackalopes around here." The woman smiles. "We're proud of 'em. I ought to talk to Bill about getting one up in here. I'm sure Jim Herrick would make us one. Are you gonna see Jim while you're here?" she asks.

"I talked with him on the phone. What a nice man. He said he'd have been glad to meet me, but he'll be out hunting all week. He gave me good leads to some other folks I should talk to, though." Then, before wrapping up our conversation, I ask the young bartender one last question: "Why do you think people love jackalopes?"

She pauses, clearly taking my question seriously. "I guess it just brings a little joy into your life, you know? Sometimes silly things are just what the doctor ordered. You never know when you'll need that little good feeling you get whenever you see one. You pretty much have to smile when you see a jackalope."

I nod and thank her. She hands me the room key and another IPA and I make my way back into the lobby under the watchful eyes of the bison mount and up the creaky wooden stairs to my room. I'd forgotten the satisfying feeling of turning a hotel room door lock with an actual key. Inside the room—the floor of which is noticeably sloped—are two small beds placed at odd angles and facing each other in opposite directions. I choose the one farthest from the noisy heating unit and flop down. A thousand-mile drive and no jackalope mount on display in the historic LaBonte bar, but at least I'm on the trail now.

The next morning I awake at dawn from a strange and wonderful dream. In it I'm standing alone, on a high ridge, gazing out over an unbounded shortgrass prairie that is being grazed by a vast herd of antlered rabbits. In the dream there is no sense of the alien—no feeling of anything strange or unusual—and the exhilaration I feel is identical to the sensation I experienced the previous afternoon when I spotted a large band of pronghorn wandering a riparian meadow adjacent to the highway. The dream scene is as sublime as an Albert Bierstadt western landscape painting: the endless prairie, broken only by a few rocky buttes and dotted with hundreds of placid jackalopes, rolling beneath effulgent clouds to an impossibly distant azure horizon.

I spend the early morning in my crooked hotel room preparing notes for the day's interviews in my journal. When I finally leave the room and walk down the narrow hallway, I notice that the old elevator has only an UP arrow as an option. Next to the button is a typewritten sign that reads, typo and all, "PLASE USE 'UP' BUTTON TO GO DOWN." I decide instead to descend the wooden stairs, which bring me to the hotel desk where a neatly dressed middle-aged man with a black mustache sits reading the newspaper.

"Good morning! You must be Mike? I'm Bill Kalar. Welcome to the LaBonte! Did you have any trouble last night? Was everything OK? And, most important, did they tell you about the hotel?" he asks, cheerfully.

"All good. Thanks, Bill. About the hotel? You mean about the history of this place?"

"No, no. About the haunting. This place is haunted! Has been for a long time, so we're used to it, but we always like to warn visitors," he says, with a smile that is warm and guileless. "Over a hundred years old, this place. Plenty of time for stories. *Plenty of time for ghosts*," he whispers, as if to avoid rousing the resident spirits.

I nod and smile awkwardly, thank him again, and head out of the hotel as he calls after me, "Mike, you have yourself a great day!"

My first stop is the Wyoming Pioneer Museum, adjacent to the state fairgrounds. The walk is short and the town feels sleepy as I pass old-style streetlights festooned with sky-blue banners proclaiming "Jackalope City" and adorned with the unmistakable silhouette of the antlered bunny. I soon arrive at the museum, whose entrance is marked by a beautiful, life-size bronze statue of two pronghorn. There I am met by the curator, Jenna Thorburn.

"Very nice to meet you, Mr. Branch. Thanks for visiting us. I've pulled all the jackalope-related materials from our archives for you. But maybe you'd like to begin by seeing our public display?" Thorburn guides me past the gift shop, introduces me to a few of her colleagues, and then we enter through an open door into a vast gallery of carefully arranged displays. The walls are graced with framed maps and documents relating to pioneer days in the Wyoming territory. Glass showcases display old cowboy accoutrements: custom saddles, spurs, bits, bridles, and blankets, as well as period pistols and rifles, vests, boots, and hats. A few cases exhibit meticulously hand-sewn, white wedding dresses dating to the 1880s. One corner of the museum is outfitted to represent an authentic frontier cabin kitchen, complete with a wood-fired cook stove, hanging oil lamps, and cast-iron skillets. Another corner has been arranged to recreate a small parlor, featuring a Victrola, foot-pumped bellows organ, and antique sheet music.

I turn to face the wall through which we have entered the gallery. On it hangs an immense bison head—the furry, brown noggin of an animal that in life would have weighed more than a ton—flanked by trophy elk mounts, each bearing antlers spanning more than a yard. To the right of these monumental specimens of Great Plains wildlife is a display consisting of three little jackalopes: two impressive, wall-hung shoulder mounts and a beautifully executed full-body mount. I approach the display with a strange reverence; this is my first opportunity to observe a mount made by one of the jackalope's creators. This is as close as I'll ever get to Doug Herrick's original jackalope, which for decades hung in the bar of the old LaBonte.

"Luke and his wife Grace loaned these to us for a temporary display, but visitors love the jackalopes so much that we decided to leave the exhibit up indefinitely," explains the curator.

A placard accompanying the display written by Luke Herrick, a friendly guy I had exchanged phone messages with before my trip, is titled "The Last Jackalope." Part of Luke's testimonial reads:

> As I sit at my home in Douglas, WY I glance at the shelf that holds my Grandpa's (Ralph Herrick) last Jackalope he ever made. He had went to my dad (Jim Herrick) at the age of 81 years and told him he would like to make one more last Jackalope. So they both went out of town and shot a jackrabbit and proceeded to make the mannequin, which is the form under the hide. Grandpa gathered all his tools together and proceeded to make his full mount Jackalope. We were all proud to know that he had once again shown off his wonderful ability to provide us with a great piece of traditional art and humorous personality. . . . The legend of the Jackalope is known far and wide around the country and of course in Douglas, WY (The Home of the Jackalope). And I am proud to let others know that the tradition of making the Jackalope and taxidermy lives on for three generations in the Herrick family. And it all started with two young boys with humorous personalities out on a homestead outside Douglas, WY: Douglas Herrick, the inventor of the Jackalope, and my Grandpa Ralph Herrick who continued to make the Jackalope and wouldn't let the legend die; a tradition still being carried on today by both families.

Standing before a vintage Ralph Herrick jackalope full-body mount, I begin first to smile and then to laugh aloud. I have been looking closely at jackalopes for years, but never have I experienced a creature quite like this one. Many jackalopes stare past you, glassy-eyed, their lifeless gaze affirming their ironic status as pure kitsch. But this antlered rabbit wore a sardonic smirk, and his posture and eyes were expertly fabricated to project serious attitude. This little guy looked not at but rather *through* you. There was an insouciance, a kind of quiet, offended dignity about this Herrick jackalope—a sense that while you might find him comical, he most certainly did not share your view. It's that look of decorous affront that made the hybrid beast so charming and funny. To gain access to the museum's archives I had playfully billed myself as a "serious jackalope researcher." Now I was standing before an authentic Herrick jackalope, chuckling uncontrollably.

"Yes," says Thorburn, smiling. "They do have that effect on people."

After regaining my composure, I am led to a back room of the museum, where I'm left alone with a single box of jackalope artifacts, which the curator lays before me as if it is the Magna Carta. I respectfully don the white cotton gloves used by archival researchers and begin my perusal. There are some documents related to early proponents of the jackalope, including Jack Ward, a member of the Douglas Chamber of Commerce who energetically promoted the beast from after World War II until the early 1960s. And Red Fenwick, a newspaperman out of Cheyenne whose often-humorous *Denver Post* column pumped out stories that spread the jackalope's fame beyond Wyoming's borders. There are jackalope pins that look as if they date to the 1960s and '70s, and also handwritten notes associated with the Herrick jackalope display: who in town still has a first-generation Herrick mount, who could loan one to the museum, who might remember the story of what happened to that first jackalope which once hung in the historic LaBonte bar. And there are wonderful postcards, some dating to the '50s: a giant jackalope on a ridgetop, dramatically backlit by the setting sun; a jackalope saddled up and being ridden by a bronco buster; a hilarious trio of amorous jackalopes cavorting weirdly on their hind legs amid the crashing bolts of a lightning storm.

> NOW, THEREFORE, I, ED HERSCHLER, Governor of the State of Wyoming, do hereby proclaim Douglas, Wyoming to be the official
>
> HOME OF THE JACKALOPE
>
> and I urge all residents of Wyoming and other states to recognize Douglas, Wyoming, as "Home of the Jackalope." I encourage those writers and organizations which have attempted to rustle the origin and home of the Jackalope away from Douglas, Wyoming to cease and desist from this misbegotten activity.

On my way out, I pause to take a last long look at Ralph Herrick's wonderfully wry jackalope, and I laugh again. I thank Thorburn for making the museum's jackalope archives available to me. Before I turn to leave, I ask her my enduring question: "Why do you think people love jackalopes?"

She chuckles softly. "You can believe it would be possible even though you know in your heart that it can't be."

My next appointment is with Bruce Jones, Mayor of Douglas. After walking back to the LaBonte I drive a few miles to the edge of town, where I'm to meet the mayor at the restaurant of the local golf course, which is called Douglas Golf Course and is located on Golf Course Road. Aside from jackalopes, people and things are refreshingly straightforward around here.

The restaurant is a wonderfully unpretentious place, more like an old-school soda fountain than a martini-mixing nineteenth hole. Having neglected to dig up a photo of the mayor, I don't have the slightest idea who I'm looking for. But Bruce Jones knows everybody in his town so he immediately recognizes me by virtue of not recognizing me. Jones is a big man with a warm smile, a goatee, and a tattoo on his upper arm. He introduces his wife, Cathy, who is friendly and a bit reserved. I try to put myself in their shoes: they are representatives of the town they grew up in, love, and serve, and they are now meeting a wild-eyed, obsessed guy who journeyed one thousand miles from the high desert to interrogate them about antlered bunnies. Who would be hospitable to somebody like that? The Joneses, as it turns out. They couldn't be more welcoming, and our conversation is unforced and illuminating. Within ten minutes I'm determined to reelect Mayor Jones, and I don't even live here.

The restaurant's only waitress brings me an iced tea with a wedge of lemon as I ask Jones about the jackalope in the context of his role as mayor.

"Well, we're the home of the jackalope. So, part of my job is to remind people that there's only one of those. A lot of folks have tried to claim otherwise over the years. They put up a jackalope over in Dubois—that's a couple hundred miles west, out toward Jackson—and said it was the biggest jackalope anywhere. Not true. The statue here in Jackalope Square is much bigger," the mayor says, with obvious pride. "There's always somebody claiming they invented the jackalope. Once the Governor of Texas tried to make the jackalope his state's official mythological creature. Well, I put a stop to that. We've had trademark on the jackalope since '65." Jones is polite, but unwavering in his conviction that Texas ought to stick with the chupacabra.

The mayor goes on to tell me how much fun he has with the jackalope myth, especially when he's visiting Washington, D.C. on business. Since colonial times it has been an American tradition for folks from the hinterlands to dupe city slickers whenever possible, and the jackalope has performed reliable service as a hoax. Jones is fond of telling representatives from other states about the animal, showing them pictures of it, and inviting them to visit rural Wyoming to join him for a little jackalope hunting. He insists he has reeled in plenty of credulous politicians with these stories.

"It's like the snipe hunt," he says, referring to a gag that has been going strong since before the Civil War. "Jackalope hunting is the same idea. It's D.C. They don't know anything about Wyoming. 'Do you really hunt them?' is what they always ask me. *Well, sure we do!*" he says, dropping effortlessly into a lilting, rustic twang. "We're a whole town with an inside joke. Everybody goes along with it."

Bruce's wife, Cathy, who has been nodding and smiling, enters the conversation with a fresh take on the horned bunny.

"When I was a little girl, I thought every town had a jackalope. I remember when my cousin told me it wasn't real and I didn't believe him, so I asked my grandma, and she had to tell me. That was so hard! Here in Douglas, Jackalope was like our Tooth Fairy. We'd leave out a bag of raw oats for him, and if the oats were gone in the morning, it meant you'd have good luck. I think I believed in Jackalope longer than I believed in Santa Claus. For her first birthday, I gave my granddaughter a pair of jackalope booties that I knit." Looking genuinely wistful, Cathy looks down and sips her iced tea. Bruce smiles at her affectionately.

I ask Bruce about the branding value of the jackalope to the town he represents. He replies that in the eight years he's been mayor there hasn't been much pushback against using the jackalope to market Douglas. The town hosts the state fair, he says, but that's just once a year. And they have plenty of clean coal but can't get a coal port, he adds. The current gas and oil boom will help, but the beauty of the jackalope is that it hasn't been a boom-and-bust phenomenon, but rather a reliable, recession-proof means of keeping his little town on the map. Jones pulls out his phone, turning its screen toward me.

"Here's our brand new logo and motto," he says with a laugh. "Some of the older folks don't care for it much, but the younger ones like it. Jackalope is just like a town: you've got to keep it alive as you go." On the small screen is a modern-looking, stylized image of pyramidal mountains and, integrated into the range, the sweeping, dynamic, impressionistic form of a leaping jackalope. Beneath the logo is the new slogan Bruce has developed to re-envision his town's mythical creature: "City of Douglas, Wyoming. Home of the Jackalope. *We know Jack.*" I smile and tell Jones how much I like it. He laughs again.

Before thanking the Joneses for their hospitality, I ask my question, first of the mayor.

"Bruce, why do you think people love jackalopes?"

"Wyoming is the least populous state in the lower forty-eight," he replies without hesitation. "Six people per square mile, and I can remember when it was half that. There could be anything out there. Who's to say it can't happen? Everybody's heard of a labradoodle. Would you believe me if I told you we have a pet wild antelope named Jake who has sixteen-inch prongs and grazes in our backyard? Well, we do."

Now it's Cathy's turn.

"I guess it goes back to Santa and the Tooth Fairy. It just kind of gets into your head. It's a piece of innocence you never have to lose, and it stays with you even after you find out. Every fable has a grain of truth to it."

My final stop is the town's visitor center, which is set up in a charming old train depot near the North Platte River, a broad, shallow, gleaming beauty that meanders languidly through town and beneath Jackalope Bridge. To a person, everyone I've talked to and corresponded with in preparation for this pilgrimage—including members of the Herrick family—has recommended that I meet with "the jackalope lady," Helga Bull. Given the jackalope's origin as a hoax, I speculate that Bull is as much a myth as the jackalope itself or that, if real, she has the perfect surname for a teller of jackalope tales.

The very real Helga Bull is expecting me and lights up as I enter the old building. The jackalope lady is pure warmth, energy, and enthusiasm, and it's immediately clear that she's in the right line of work. Before I can ask my first question, she launches into what I can tell is her shtick, but it's such an enjoyable one that I have no urge to preempt it.

"Here's what you came for, Mike!" Bull exclaims, gesturing toward a wall featuring three jackalope mounts with a fourth full-body mount displayed on a table below and, to the right of and below the table, a fifth jackalope perched upright in a cage. The wall also features a bright yellow "Jackalope Crossing" sign, an original of the horned rabbit medallion that decorates the nearby bridge over the North Platte River, and a large framed replica of Governor Herschler's proclamation officially designating Douglas the Home of the Jackalope.

"Now, there are two species of jackalope. These here, with the branching antlers, are the mountain species. But sometimes you see the plains species, which has horns like an antelope. Of course, these are all bucks. The doe jackalope doesn't have horns or antlers. Both the mountain and plains species are related to a much larger jackalope species called the saber-toothed jackalope that's probably extinct now. Those ones weighed up to 150 pounds. Bones of that big fella have been found all over Converse County!"

As Helga Bull continues her jackalope primer, I notice that one of the creatures on display is wearing a blue bandana tied loosely around its neck. A sign posted near the caged jackalope, which has been placed at kids' eye level, warns younger visitors not to stick their fingers into the cage because "Slick" might bite.

"Old timers around here said those big ones sometimes attacked wagon trains or homesteads, but nobody knows for sure. We do know that jackalopes roam together in groups of ten to twenty individuals. Scientists call this kind of group a 'committee' of jackalopes. In the springtime . . ." Helga continues her animated and impressively detailed lecture on the natural history of the jackalope, covering information about the creature's habitat, mating, feeding, shelter, speed, camouflage, predation, herding, evolution, thermoregulation, recent sightings, and threats to the species. Wrapping up her enthusiastic introduction to the imaginary animal we both love, she adds, "But then you said you're a serious jackalope researcher, Mike, so you must know all that." The jackalope lady smiles at me and I smile back. As a fellow aficionado, I recognize the twinkle in her eye.

But I do need some real answers, so it is time to coax Helga Bull out of character. I ask if the jackalopes in her display are Herrick mounts. She confirms that they are, but clarifies that the mounts were made not by Ralph Herrick—as were those at the Pioneer Museum—but rather by Ralph's son, Jim. Jim's creations bear a family resemblance to his dad's, while also having a signature quality that is entirely his own. His critters are a little gentler, somehow more dignified, more aware of their elevated status as a trophy mount, a little less ornery and sarcastic. I ask her if it is true that, as I've heard from others, Jim has made tens of thousands of jackalopes since learning the craft from his father and his Uncle Doug. "It wouldn't surprise me," she replies. "I think he's been making them almost his whole life."

A man comes through the back door and introduces himself as Bill—not chipper Bill Kalar of the haunted LaBonte, but Bill Truitt, the native of Douglas who works with Helga at the visitor center. At that moment I happen to be asking about Jackalope Days, a festival the town holds each year to celebrate its iconic mascot.

"Oh, we have a lot of fun!" Helga says. "We have a car show, and lots of crafts, and a food truck. There's a contest where you can enter your own homemade beer. One of our local businessmen dresses up like a big jackalope and gives all the kids candy. They love it! And we have a street dance too. Oh, and the dodgeball tournament! Everybody comes out for it."

Now Truitt chimes in. "I remember the very first one, back in '84, when I was a kid."

"Was that the one where somebody got loaded and plowed down the big jackalope statue that had been out on the median of old Yellowstone Highway since the sixties?" I ask, referring to the main road thorough town.

"Yeah," Bill says, chuckling. "When our local deputy showed up, that fella climbed out of his pickup and swore that the big jackalope had jumped out in front of him." He laughs again.

"That was May 19, 1984, when the eight-foot jackalope got hit," says Helga. "But on May 18, 1985, we put up another jackalope, and that one was even bigger than the first one! But after the old town grocery burned down and the family gave the land to the town, we made Jackalope Square and moved the big jackalope over there. Later on some people saw that big jackalope and wanted to use our town to do a jackalope TV show that would be like that show *Hunting Bigfoot*. We said no way!"

"No way," Bill affirms. "Something like that would make us look like a bunch of idiots."

"What kind of responses do you get from visitors who don't already know about the jackalope?" I ask.

"They love it!" Helga says. "Especially the kids. But grownups too. I'd say about half of them leave believing they're real. And that's okay."

My mind flashes to my writing desk at home, where I keep pens in a jackalope coffee mug emblazoned with the reassuring slogan: "The Important Thing Is That I Believe In Myself."

"Yup, people love them. I always ask visitors if they've seen any jackalopes on their drive in, and of course they've always seen jackrabbits, so I tell them that those are the jackalope does. And then they point to these display mounts and ask me 'Are they real?' And I say, 'Yes, they are.' And they *are* real, Mike. They're real jackalopes!" Helga's answer has me smiling.

"Why do you think people love jackalopes?" I finally ask, looking first at Bill.

"The jackalope is kind of like that giant ball of string they got over in Kansas. Just an oddball thing. I think that's why," he says.

Helga Bull, beaming, has a ready answer. "Because they're as real as you want them to be."

I thank them both and gather up my journal, baseball cap, and travel coffee mug to head for the door.

"Before you go, Mike, the town of Douglas has a gift for you. We think you're going to need this. You've come a long way to be with us here in the Home of the Jackalope, and we appreciate that. Now, you take good care of yourself." Smiling, she hands me a sealed envelope and shows me politely to the door. More thanks are exchanged, and when I offer Helga Bull a handshake, she hugs me instead. Like everyone I've met in Douglas, Helga has a way of making me feel at home.

Outside it's brisk and sunny, and I pause to take photos of the visitor center's ridiculously large jackalope statue. It is painted entirely white, with black horns (not antlers, so it must be the plains species). It has a placid look on its face, expressing a kind of serenity that is cute but lifeless when compared with the Herrick family mounts. Even more impressive than the statue is the strange shadow it casts in the late afternoon sun. In the silhouette of that dark shadow, the giant creature appears to be walking upright with an oddly hunched back, and the outline of its huge ears and gracefully curved horns etches the ground beautifully. I take a picture of the giant shadow jackalope with my own full-body shadow standing next to it. Then I step in front of the statue, causing my own shadow to vanish, and photograph the shadow jackalope again. No one who later sees this striking image will think to ask why the photographer's shadow is not also cast. The work of the imagination is *real* work. Maybe the jackalope is best understood by the shadow it casts, even when our own shadow is hidden within it.

Tonight will hold more baseball, IPA, and good conversation with ranch hands, hunters, and oil patch roughnecks in the bar of the LaBonte. Tomorrow at daybreak I will drive across Jackalope Bridge, catch a glimpse of the giant jackalope art installation perched on a nearby hillside, and then roll a thousand miles home to the high desert. For now, I relish sitting quietly in the sun-warmed cab of my truck, sipping lukewarm coffee, and scribbling notes in my journal. When I unseal the envelope Helga Bull has given me, I find within it a piece of authentic Americana: a Limited, Non-Resident, Jackalope Hunting License, duly issued to me by Converse County, Wyoming, and the town of Douglas. The document offers a brief history of *Pedigres Leapusalopus*, which was reportedly first sighted by trapper Roy Ball, who "staggered into the Wyoming territory around 1829." "The existence of the jackalope was known prior to that only through Native American legend," the account continues. "Mr. Ball was the first to record that jackalopes mated during flashes of lightning common to violent thunderstorms of the prairies. This, however, is yet to be confirmed by modern zoologists." Under the section "Legal Notice and Regulations," the license specifies that "The Wyoming Fish and Game Commission requires that jackalopes be hunted only between sun-up and sun-down on June 31 of each year."

I turn the paper over and read the official language aloud in order to formally take the oath: "I, Michael P. Branch, being first duly sworn, do hereby state: That I am a person of strict temperance and absolute trustfulness. I do reserve the right, at my discretion and if interrogated concerning my hunting experience, to employ such lingual evasion, loud rebuttal and double talk as the occasion and circumstances require." The license is signed by the invented Chief Licensor, Adam Lyre, and by the real First Lady of Jackalopes, Helga Bull.

Well, it's official, I think to myself. *I am in now in possession of a duly sworn legal warrant authorizing me, as licensee, to hunt, pursue, trap, or otherwise go after the jackalope in any way I can possibly imagine.*

And that is exactly what I intend to do.

Chapter 2
Believing is Seeing

As adults, anyone who likes to tell stories harbors the secret desire to one day pull off a real whopper. Not to be ornery, or mean, but just to stand tall. To feel good. To take revenge on the everyday and the ordinary.

—Barry Sanders, "How to Tell a Story in America (Make It All True, Damned Near)"

Have I ever told you about the time I took my daughters, Hannah and Caroline, out into the wild Nevada outback on an epic day trip? There are all kinds of lies told about this vast high desert—stories cut from whole cloth by old buckaroos and pocket hunters and rounders and passed along from one desert rat to another. But I've spent decades exploring this wild country, studying its geology, wildflowers, animals, and weather, and I've seen some incredible things. The Great Basin Desert is the largest, highest, coldest, most isolated desert in North America. On the day I'm telling you about, I took the girls out to Pershing County, north of the little desert oasis of Lovelock. This county is about four times the size of Rhode Island and has a population density of 1.1 people per square mile. Even during the nineteenth century, two people per square mile was the official U.S. Census Bureau standard for "unsettled" frontier. This Great

Basin wilderness survives outside time, as vast and unpeopled as it was 150 years ago. We call this magical space "the Big Empty" and know that anything can happen here.

The girls and I loaded up Alkali, my old pickup, with some food, the map and compass, and basic emergency equipment: snakebite kit, emergency signal mirror, seven gallons of water, a bottle of High West Son of Bourye whiskey, and a cooler containing apple juice for the girls and a crowler of Revision Dr. Lupulin Triple IPA for me. We love this wide open land, but if your truck breaks down in a remote canyon or on a desolate playa, you could be done for. I once discovered a human skeleton while solo hiking in the Seven Troughs Range. It had an old compass in one of its bleached, bony hands; an empty water bottle lay near the other. Stories about mountain lion attacks are mostly exaggerated, though, and the scorpions are huge and hairy but don't sting very often. It's freezing in winter, blazing in summer, and has extreme day-night temperature swings. With the exception of a few well-hidden springs, no water is to be found.

The girls and I followed I-80, winding along the Truckee River, past the historic Mustang Ranch brothel, out beyond the little Paiute town of Wadsworth, and then further east through the waterless alkali wasteland of the forty-mile desert, the most dreaded and grave-littered stretch of the old California Trail. I pointed out to the girls an expanse of highway where I once saw Mormon crickets (actually shield-backed katydids) moving across the land in swarms so thick that, after the mining haul trucks rolled through, the insects were squished deep enough that the highway became slick as ice from mashed cricket guts and had to be closed by the highway patrol.

North of Lovelock, where the country opens up, I took the girls out to a playa where I like to enjoy Nevada's signature form of outdoor recreation. First, drive onto the playa, an enormous, bright, white, baked, cracked, flat, disorienting expanse of hardpan desert that is the most isotropic landscape on the planet. Next, jimmy your gas pedal. Finally, climb out your window and onto the roof (I recommend taking a few cold beers with you) and let your unmanned truck drive itself wherever it happens to go. That's how vast and uniformly flat these playas are. The day I'm telling you about I even let Caroline drive on the playa, but since she's only twelve years old, I don't allow her to go above 80 mph. She worried that she might drive us at high speed straight into the immense lake—until I assured her that the "lake" was nothing more than a shimmering mirage. Caused by refracted light rays dancing along the surface of the overheated playa, the shoreline receded as she sped us toward it.

Leaving the playa and continuing overland into the sage-dotted hills, we pulled over at the base of a rock-strewn desert mountain and set out hiking through what looked like pretty good jackalope habitat. We spotted a prairie falcon guarding its nest on a ledge a hundred feet up in the cliffs—a nest it probably took from the ravens—and a soaring golden eagle tracing lazy circles on a big thermal, and a towering V formation of white pelicans, which have a wingspan of around ten feet. We identified four species of lizards but unfortunately saw no desert horned lizards, which can spurt blood out of their eyes. There was horse dung everywhere, and from a high ridge we gained a fine view of nine wild mustangs wandering the draw below, their manes bristling in the warm breeze. In a desiccated arroyo we discovered a prospector's rusted-out gold pan and plenty of lithic scatter, good-sized chips of chert and obsidian flaked from arrowheads and spear points, in colors ranging from a glossy, volcanic black to buckskin tan and sunset crimson. Since this desert was an ocean 200 million years ago, we also hunted for the fossilized bones of ichthyosaurs—sea lizards that reached 50 feet in length—though we weren't lucky enough to find any. In the dense sage and rabbitbrush we scared up a few desert cottontails (*Sylvilagus audubonii*) and black-tailed jackrabbits (*Lepus californicus*). The jacks can run 30 mph and jump 20 feet, and they dash a getaway pattern as zigzagged as a bolt of lightning streaking across the ground.

From a distance we spotted a band of pronghorn on a far hillside. Often mistakenly called "antelope," the pronghorn (*Antilocapra americana*) is a one-of-a-kind beast, nearly as fantastic as a unicorn. While *Antilocapra* means "antelope-goat," the animal is neither. A unique species, the pronghorn evolved here in North America over the past twenty million years, becoming an incredibly fast runner in order to escape its main predator, the American cheetah, which vanished about 10,000 years ago. Some people say pronghorn can run 90 mph, but that's a folktale used to fool greenhorns. They can only run about 60 mph, which still makes them North America's fastest land animal by a wide margin.

The thing I should have told you before—but didn't because I was distracted by the falcons and pelicans and lizards and mustangs and pronghorn—is that we saw at least one hundred wild burros, and maybe closer to two hundred. They've been out in this desert since the Spaniards brought them here in the 1500s, so they're at home by now. Most of the jacks (males) were standing solo, while a lot of the jennies (females) stood side-by-side with their cute little burritos (first-year foals). Folks are often fooled by the burro's sweet appearance, but the animal can be ferocious when frightened. That's why I hesitated when Caroline asked me if she could ride one. But she's a pretty tough twelve-year-old, and so, thinking it would be a memorable experience for her, I agreed.

We hiked back to Alkali for snacks and apple juice for the girls and whiskey and beer for me. As we rested, I did my best to teach Caroline exactly how to ride a wild burro. As I explained, the trick is, first, to put the animal into a spell by talking to it very sweetly. For example, you never want to disrespect a burro by calling it a mule or donkey, let alone a jackass. Then you want to ease up to the burro slowly and gently stroke its neck. If the animal isn't too skittish, that petting will make it tranquil, and that's your now-or-never moment. The final move is to grab fast to the animal's mane with both fists and swing yourself up onto its back. As any Nevadan who has spent time out in the Big Empty will tell you, that's really all there is to it.

Caroline said she was ready to give it a try, so I took a few swigs of Bourye, washed it down with some Dr. Lupulin, and we headed out toward a small scattering of burros about a hundred yards away. Caroline selected a burro that was a lovely shade of white, easing up to it exactly as I had told her to. I stood at a distance, encouraging her in a soft voice so as not to spook the animal. Hannah was beside me, quietly strumming a song on her Little Martin travel guitar. Music is soothing to burros, so Hannah plucked a tune called "Wild Burro of the West." While we kept our distance, Caroline approached him very slowly, whispering nice things like. "Mister Burro, you sure are a pretty color," and "Boy, I could get you a straw hat to make some shade for your noggin." She got right up to him, and as she started stroking his neck he began to look mighty relaxed. Well, next came a long pause, and I could tell that Caroline was getting ready to grip the white burro's shaggy mane.

In a quick second everything exploded like dynamite down a mine-shaft! Caroline grabbed that burro's mane and nimbly swung herself up onto his back, and in the same instant they took off like a shot. I lit out running after them both, with Hannah running along behind me, frantically strumming in hopes of calming the animal down. Caroline had her cheek down low on the side of the white burro's neck holding on for dear life. Her braids were flying straight back and the burro was running almost as fast as a pronghorn being chased by a cheetah, heading for a part of the desert that was choked with boulders as big as trucks. I worried he might be thinking about scraping Caroline off against one of those huge rocks.

I knew there would be real trouble if the burro slammed my little girl against those boulders, but the sun was blazing, and the ground was sandy, and the whiskey and beer had slowed me down a little. As they streaked away I didn't see any way to reach the rocks before that white burro did. But just as the burro, in full flight, ran for the boulders, it stepped on a Great Basin rattlesnake, and that big coiled buzzworm struck at the front legs of the burro, fangs out, missing its strike but spooking the burro and causing it to lurch. It pitched Caroline right off its back and she went flying like a kid being thrown over the handlebars of a bicycle. I could see right away that she was fine, but now we had a more serious problem. That burro was good and mad, and he stamped his front hoof twice to announce his intentions before starting to close in on her. The big rattler, which was even more furious than the burro, slithered ahead of the burro directly toward her. As both vicious animals descended on my defenseless daughter, I scrambled for a plan. But I was unarmed, still 50 feet away, and Hannah's song wasn't helping one bit.

In that moment of impending doom, something shot out from behind the big boulder closest to Caroline and leapt over her in a spectacular jump of at least 30 feet. At first the animal looked like a jackrabbit, but when it landed I could clearly make out, alongside its tall ears, a pair of branching sharpened horns, backlit and gleaming in the high desert sun. It was a jackalope! The jackalope is often called the "Warrior Rabbit" because its speed, strength, sharp claws, teeth, horns (technically antlers), and terrible disposition make it capable of incredible violence. I was terrified for Caroline, as I knew all too well that the Warrior Rabbit can eviscerate humans by goring them with its formidable horns.

The next moment was as extraordinary as any I've ever experienced in this vast, wild desert. Just as the rattler was about to reach Caroline, with the angry burro stamping along close behind it, the jackalope pounced on the snake in a blinding attack, burying its sharp teeth into the buzzworm's neck just behind its triangular, venom-laden jaws. Then the jackalope slung the snake with such blinding force that the rattler wrapped around the legs of the burro, bringing it tumbling to the ground. As the burro lay on its back, flailing its hooves in the air, the jackalope leapt onto the burro's belly and finished the job by grabbing the rattler in its paws and using the live snake to tie the burro's hind legs together, just like a cowboy tying down a roped calf. Having knotted up the wild burro with the rattler, the jackalope darted behind the boulder. I reached Caroline in the next moment, helped her up, and we all walked back to Alkali, where we had more snacks and drinks before setting out for home.

The jackalope, whose incredible attack saved my daughter's life, was not the American Jackalope (*Lepus cornutus americanum*), which is rare in the Great Basin Desert, but instead the more common Alkali Area Jackalope, which the *Field Guide to the North American Jackalope* formally identifies as the Western Jackalope (*Lepus cornutus occidentis*). Because so many stories about this remarkable animal are distorted and inaccurate, I am currently writing up this extraordinary incident for the *International Journal of Behavioral Mammalogy*, whose editors are keen to have a contribution describing an animal so little known and so poorly understood as the extraordinary jackalope.

The story I have just fabricated for your entertainment is what folklorists call a "tall tale," and what storytellers in various regions of the US call a "yarn," "stretcher," "windy," or "whopper." These kinds of stories are as old as the republic, having been created by Americans as an inside joke to fool European listeners or readers who had no direct experience of the fabled American land. Even our esteemed Benjamin Franklin loved to pull the wool over the eyes of the uninitiated with hyperbolic narratives. In what is among our country's first and finest "fish stories," for example, Franklin wrote in 1765 that "whales, when they have a mind to eat cod, pursue them wherever they fly; and that the grand leap of the whale in the chase up the Falls of Niagara is esteemed by all who have seen it, as one of the finest spectacles in nature." How many readers of the London *Public Advertiser* believed that a whale could leap 167 feet up a waterfall (never mind the fresh water!) must remain unknown, though Franklin's reputation as a pioneering scientist and his rhetorical skill in weaving colossal lies seamlessly into a rich fabric of credible details certainly allowed him to reel in plenty of credulous readers.

Like Franklin's delightfully improbable story of the leaping river whale, most tall tales and myths, including those that emerged from Native American and African American traditions, express and amplify aspects of the American land. "[T]he tall tale has held a place of special significance in American life," explains Carolyn S. Brown in her study of the form. "From almost the beginning, the incomprehensible vastness of the continent, the extraordinary fertility of the land, and the variety of natural peculiarities inspired a humor of extravagance and exaggeration." As Barre Toelken put it in "Folklore in the American West,"

> Our main aim is not to make sober distinctions between truth and falsehood in western culture; to begin with, truth is too hard to capture. How would anyone describe the redwoods or the canyons, or the colors, or the heat/cold extremes for someone who had never experienced them? Indeed, the truth is hard enough to deal with: in Oregon, if an angler catches a sturgeon under three feet long or over six, it must be returned to the water. Explain *that* one without smiling to your relatives in Connecticut.

Just as Franklin's far-fetched stories functioned as an inside joke at the expense of duped Europeans—one reason the tales were treasured by Americans—by the nineteenth century, the tall tale was a narrative form widely deployed by rural Americans as an inside joke at the expense of city slickers, the urban (and, usually, eastern) elite whose condescension to frontiersmen made them a perennial target of this kind of humor. During this same period the tall story was also used by African American tale tellers to satirize the unjust power exercised by the dominant white culture—a potent form of resistance evident in narratives ranging from simple Br'er Rabbit tales all the way to the sophisticated literary performances of Charles Chesnutt's *The Conjure Woman* (1899). Many Indigenous North American myths likewise figure trickster characters—including coyote, raven, mink, and blue jay—whose extravagant antics and hyperbolic language often expose the violence of the colonial project which decimated native cultures.

Although the tall tale is an unusually malleable narrative form, it is characterized by a number of defining elements, or genre conventions, which I've constructed my own tall tale to illustrate. Many tall tales open with a condemnation of mendacity, as when I decry the fact that "There are all kinds of lies told about this vast high desert—stories cut from whole cloth by old buckaroos and pocket hunters and rounders and passed along from one desert rat to another." Early in a tall tale the narrator works to establish their authority, as when I stake the claim that "I've spent decades exploring this wild country, studying its geology, wildflowers, animals, and weather. . . ." Like Franklin, I take pains to include rich detail, relating facts about the geography and animals of the high desert, with a special emphasis on truths that might appear incredible to an uninitiated reader. Is the population density of the Great Basin really that low? (Yes.) Did Mormon cricket guts actually prompt the closing of a highway? (Yep.) Can it be true that playas are so vast and flat that you can sit atop your vehicle while it drives itself randomly across the desert? (Absolutely. In fact, you can see me doing this very thing in the video trailer for my book *How to Cuss in Western*.) Do American white pelicans really have a wingspan of almost 10 feet? (You should see them!) Can horned lizards actually spurt blood out of their eyes? (You bet. Watch out!) Can jackrabbits run 30 mph and jump 20 feet? (It's a thrill to witness!) Did we really discover an old miner's rusted gold pan, and find pieces of arrowheads, and see wild mustangs, and hunt for Ichthyosaur fossils because this desert was once ocean? (We sure did.) Is the pronghorn actually a unique species that can run 60 mph because it evolved to escape from now-extinct cheetahs? (Undeniably.) By including so many improbable yet factual details, I work to establish a narrative authority that will help me string readers along. Note, also, my use of scientific names for many of the species I mention. As Franklin knew, the fact that scientific discourse sounds unimpeachably authoritative makes it an effective smokescreen for the double-dealing tale-teller.

Of course it isn't the truth—however implausible that truth might seem—that makes the tale tall, but rather the lie. The key to effective tall tale telling is incrementalism: the progressive addition of material that slowly strains the credibility of the narrator's story. Yes, Nevada is incredibly remote, dry, and sparsely populated, but did I really find a human skeleton while hiking? (No.) It is a fact that the male burro is called a "jack" and the female a "jenny," but is a burro foal really called a "burrito"? (Nope. It's called a "foal.") And it is true as well as incredible that my daughters and I did see more than one hundred wild burros on our day trip. But can the strange and surprising account of wild burro riding be true? (Of course not!) And yet, this false account may not seem much less credible than some of the true statements in the narrative, because the Big Empty is as otherworldly as any landscape a teller of tall tales could conjure. After all, we live in an astounding world in which birds can have ten-foot wingspans and lizards can shoot blood out of their eyes.

Unlike a con or hoax, in which the goal is strictly to deceive, a tall tale isn't constructed to be believed so much as it is crafted to entertain. I like to think of the tall tale as a tower of narrative Jenga blocks. Each fact adds a block to the tower and thus produces stability; each fib removes a block, introducing doubts about veracity into the reader's mind and causing the narrative structure to wobble. Because the incremental revelation of falsehood is the central dynamic in this sort of narrative, the real pleasure of a tall tale is the way it joyously embraces its own tallness, its own instability. By the time the jackalope is tying up the hooves of a wild burro with a rattlesnake, even the greenhorn reader can have no doubt that it isn't only the burro whose leg is being pulled. The tall tale is a narrative Jenga tower specifically designed to provide enjoyment upon its collapse.

As Mark Twain explained in his 1897 essay "How to Tell a Story," "The humorous story is told gravely; the teller does his best to conceal the fact that he even dimly suspects that there is anything funny [or false!] about it." The teller's feigned innocence—their apparent lack of awareness that the story contains any absurdities whatsoever—is part of the charm of a tall story. This is why, at the conclusion of my tale, I continue to offer superficial cover for the lie, citing the subspecies of jackalope (complete with formal taxonomic nomenclature) and explaining that I am presently preparing a scientific article about it (albeit for a journal that does not exist). As Twain knew, the teller's deadpan delivery and ingenuous pose increase the gratification that results when the story finally collapses into farce, for that collapse invites the listener to finally share the joke with its teller. The moment of revelation initiates the listener into a new sense of community with the smiling folks who know full well that, while the Nevada desert has plenty of wild burros and venomous snakes, you aren't likely to see a jackalope (of any subspecies) tie down a burro with a live rattler.

Douglas, Wyoming natives Douglas and Ralph Herrick are said to have invented the hoax horned rabbit taxidermy mount in the early 1930s. By the mid-twentieth century the jackalope had begun to attain national notoriety, as suggested by a piece called "Jackalopes Really Exist" which ran in Pennsylvania's *Altoona Mirror* on March 2, 1950. While the horned rabbit's reputation was fueled by the dissemination of mounts—and also by the plethora of joke jackalope postcards those mounts inspired—it was primarily folk narratives that spread the hybrid bunny's fame and helped it achieve legendary status. Importantly, the Herrick brothers were probably not behind the jackalope tall tales that emerged in the decades following their invention of the creature. On November 27, 1977, the *New York Times* published an article called "Wyoming, Where Deer and Jackalope Play," in which they interviewed Ralph Herrick and reported that "Mr. Herrick said he does not know how many of the legends surrounding the jackalope began, such as the one about the creature's extraordinary ability to imitate the human voice, or the story that said the first man to see the jackalope 'was a trapper named Roy Ball in 1829.'" While Herrick may not have actively promoted outrageous stories about the horned rabbit, neither was he inclined to interfere with what is often called "the folk process." "People get real mad if you tell them there's no such thing as a jackalope," Ralph told the *Times*. "They take it very seriously. And why make people mad?" You can almost picture the sly smile on the face of the jackalope's cocreator as he proffered that final rhetorical question.

But if the Herrick brothers did not actively engage in producing folk narratives about the jackalope, plenty of other folks did. Soon the jackalope emerged as a new star of the American tall tale, one head and antlers above other "fearsome critters" (a term used by folklorists to taxonomize mythical beasts) like the agropelter, dungavenhooter, gillygaloo, glawackus, gumberoo, hodag, hoot-pecker, owl-eyed ripple skipper, rubberado, sidehill gouger, snallygaster, squonk, three-tailed bavalorous, tripodero, and wampus cat, few of which are familiar to Americans today. The fun of folk narratives—in contrast, say, to Disney narratives—is that they are the product of a collective form of storytelling over which no individual or corporation has authority. The jackalope is the common property of the "folk," an expansive, amorphous group comprised of anyone who takes pleasure in the horned bunny and propagates stories about it.

As a result of this wild process of narrative germination and dissemination, many fantastic and contradictory tales are told of the jackalope. To begin with, various sources cite the animal's name as jackalope, jackelope, jackaloupe, jackalop, jack hare, jacka rabbit, saber-toothed jack, razor jack, bristled hare, deer hare, thistle hare, lion rabbit, antelabbit, deerbunny, stagbunny, sarabideer, boop-oop-o-doopdeer, whatizzitt, and warrior rabbit. Would-be scientific names for the hybrid hare have also proliferated and include *Lopegigrus lepusalopus ineptus*, *Lepus antilocapra wyomingensis*, *Leporidae smilodon*, *Pedigrus lepusalopus ineptus*, *Leporus cornuti*, *Lepus tempermentalis*, and various subspecies of the horned *Lepus* genus, including *americanum*, *occidentis*, *anglicus*, *canadensis*, and, last but also least, *minima*. At birth jackalopes are called "bunnies," unless they are called "jackalopeenies" or "leverets." Later in life they are called "calves," or, in other stories, "spikes." Adult males are usually called "jakes," while adult females are "does." A group of jackalopes may be called a "band," "pack," "jump," "flagerdoot," or, as Helga Bull would have it, "committee." No two stories represent this information in quite the same way and there is not even consensus about whether the plural of "jackalope" is "jackalope" (as with "deer"), "jackalopes" (as with "bears"), or "jackalopei" (as with "hippopotami").

Stories indicate that this unusual animal resulted from a cross between a jackrabbit and a pronghorn. Or a hare and a mule deer. Or an antelope jackrabbit and a whitetail deer. Or a jackrabbit and a now-extinct pygmy deer, desert deer, or Spanish deer. The initial cross-species mating may have been produced by two animals trapped together in a cave during a storm, though other accounts abound. One story claims that the rabbit's odd horns are the product of a genetic mutation that occurred approximately thirty million years ago. Jackalopes certainly do have antlers (in some stories, true horns), though relatively little is known about them. Some say the jackalope is the only animal that sheds only one antler until it has fully grown a replacement, after which it sheds the other. This may also be why a jackalope, ever wary, is said to enter its den walking backwards in order to use those horns to protect itself against predators.

How large is a jackalope? That, too, depends on who you ask. In his *Field Guide to the North American Jackalope*, Andy Robbins has the various subspecies weighing in at anywhere from one to ten pounds. In his weirdly exhaustive booklet *Everything You Always Wanted to Know about Jackalope*, Bill Alexander claims the animal weighs up to 60 pounds, "about the size of a medium dog" [sic]. Stories are often told of a now-extinct Saber-toothed Jack, which weighed up to 150 pounds and was ferocious, especially when it was separated from its committee and thus went rogue. According to a document circulated by the Douglas, Wyoming Chamber of Commerce (a hoax document cleverly pretending to be a reprint of a 1924 article by the non-existent science editor of the fictional *Lost Springs Daily Bugle*), "Accounts of entire settlements or towns being wiped out by these herculean hares have been generally discounted by historians, but there are reliable accounts of attacks on homesteads and small wagon trains." The demise of this immense jackalope is variously attributed to the hard winters of the 1880s, loss of habitat due to agricultural expansion, and depletion of the mammoth jackalope's main prey, the American bison. In any case, the Douglas Chamber informs us, "The loss of the giant jackalope has had many positive effects on the areas they once roamed. The buffalo has begun a comeback, barbed wire fences and windmills are now safe, and women have obtained the right to vote."

Who was the first settler to catch a glimpse of the legendary jackalope? Most tall tale-tellers hold that it was Roy Ball back in 1828, though some claim it was Wild Bill Waller, or George McLean, and a few raconteurs maintain that this first sighting was made by a frontiersman named John Colter. Whoever deserves the honor, the dates most favored seem to be 1828 and 1829. According to some tale variants it was not only trappers who claimed to have seen jackalopes, but also Buffalo Bill Cody, Jesse James, Calamity Jane, Butch Cassidy, and, just for good measure, the Sundance Kid.

It is said that long before white settlers arrived on the Great Plains, many Native American cultures were very familiar with the jackalope, and it is especially common for jackalope stories to be credited to the Oglala Sioux. The most common of these asserts that the Oglala still retell an oral narrative in which the dust devils seen out on the prairie are caused by speeding jackalopes. The fact that tall tale-tellers are never obliged to confine themselves to the plausible prompted storyteller Christopher Flynn to offer the following provocative account of the vital importance of the jackalope to Indigenous peoples:

> Stories began to surface in the late 1880s suggesting that the Seventh Cavalry, realizing they were outgunned, captured and attempted to train a number of jackalope to use against the Indians. The stories claimed the jackalopes were unleashed onto the unsuspecting natives, but were quickly turned against the Americans [sic] troops. There is some credibility to these tales, since the Oglala Sioux Indians, under the command of Crazy Horse, were documented to have had special understanding of the jackalope and had no equal in their ability to capture and manipulate the critters. Where logic has failed to explain the awful defeat at Little Big Horn, the jackalope theory does fill in the missing pieces.

General Custer's fate notwithstanding, even basic facts about the hybrid bunny remain in dispute. For example, you may wonder how fast a jackalope can run. Forty miles per hour seems to be about as low as any storyteller is willing to go, and many sources land in the 50–60 mph range. My favorite answer is that the jackalope runs 90 mph because it is a hybrid of the jackrabbit, which can do 30 mph, and the pronghorn, which can sprint an astounding 60 mph. Thirty plus 60 equals 90. Simple math! On the other hand, some maintain that the animal is so adept at camouflage that it can render itself invisible, and thus rarely needs to employ its speed, however extraordinary that speed might be.

If there is a great deal of inconsistency in the representation of the horned rabbit, that is understandable given the rarity of the animal and the wide range of witnesses who have shared their accounts. However, there are certain facts on which nearly all storytellers agree. Chief among these is the fact that jackalopes mate only during lightning storms. This odd breeding behavior not only helps account for the animal's rarity, but also supplies a magical origin for the species. Plus, tacky postcards depicting jackalopes rearing up on their hind legs and cavorting erotically amid shattering bolts of lightning are worth every penny of the 75 cents they'll set you back.

Also firm in jackalope lore are accounts of the jackalope doe's milk. Most storytellers insist that, because of the jackalope's great leaps, the milk arrives at the teat already homogenized. I have yet to find an account that deems it either safe or easy to milk a jackalope. While it is true that the creature does often sleep belly-up, they are fierce when disturbed. When Douglas Herrick's obituary ran in the *New York Times* on January 19, 2003, even the *Times* couldn't resist emphasizing this point. Referencing the jackalope kitsch available for sale in Herrick's hometown of Douglas, Wyoming, Douglas Martin of the *New York Times* wrote that "Jackalope milk is available at several stores, though its authenticity is questionable; everyone knows how dangerous it is to milk a jackalope." It is universally agreed upon that the milk has medicinal properties, and it has been credited with curing everything from bunions to baldness. More controversial is the claim that jackalope milk is a potent aphrodisiac. While enthusiastic testimonials to its effectiveness have been presented by some storytellers, evidence for this assertion appears soft. Nevertheless, a few raconteurs have argued that the jackalope's sometime nickname "the horny rabbit" is attributable not to the animal's horns, but rather to the stimulating effect it has upon those who imbibe its milk.

It is also generally understood that jackalopes can be attracted by setting out a bowl of whiskey at night. Some say a mixture of whiskey and milk is more effective, and a few jackalope hunters swear that the best bait is whiskey-soaked bologna. But there can be no question about the jackalope's fondness for whiskey—a trait universally acknowledged in the folklore surrounding the animal. Older lore holds that, once inebriated, a jackalope believes it can catch bullets in its teeth, and is thus made the hunter's prey. Most hunters agree it is dangerous to hunt jackalopes, and it's said that homesteaders once wore a length of stovepipe on each leg for protection against attacks.

It is illegal to hunt jackalopes without a license, but licenses are issued with covenants and restrictions that vary by state. Most familiar is the authorization I myself carry while tracking jackalopes: the Converse County, Wyoming license, which, under W.S. 43-1-113, entitles the licensee to bag one jackalope on June 31 between sunrise and sunset. The South Dakota permit allows hunting from June 1 to October 31, but only at elevations above 5,000 feet, and only "between the hours of midnight and 3:00 A.M., during the three nights prior to the full moon." It is not uncommon for jackalope hunting licenses to have an IQ requirement; typically, the hunter must display an IQ above 50 but not greater than 72. As Robert Lesher notes in "Quest for the Warrior Rabbit," "The season is generally restricted to a single day, and the penalties for illicit hunting or possession of any part of the animal, dead or alive, are awesome—ranging up to as much as two weeks in Casper, Wyoming, a small community to the southwest. That particular punishment, I was told in Douglas, is being contested in the Supreme Court as cruel and unusual." Although this account appears suspect, I do have it on good authority that it is considered unsportsmanlike (and, in some states, illegal) to attract the animal by imitating its mating call using a kazoo.

Arguably the most significant behavioral characteristic of the jackalope is its celebrated ability to sing, and also to "throw" its voice so as to deceive predators (including human hunters)—the only known example of ventriloquism in the animal kingdom. Many cowboys and campers have told the story of sitting around the campfire at night, crooning a ballad, only to hear the jackalope join in, often in sweet harmony with the song's melody. Most accounts specify that the jackalope sings in treble, though a few less credible tale-tellers claim to have heard jackalopes belting out the bass part. Storytellers also agree that the jackalope has the remarkable gift of imitating sounds, including those of coyotes, owls, meadowlarks, and even chainsaws. One experienced hunter swears the jackalope is capable of imitating the sound of the hunter's cell phone ringtone in order to distract him or her. Most common, though, are reports that the animal has the remarkable ability to make its voice sound as if it is coming from somewhere else. This defensive tactic is used when the jackalope is being pursued by a hunter, in which case the animal will often cry out "There he goes, over there!" in order to throw the hunter off his trail. I will add that the jackalope's use of human language as a deceptive tactic gives the lie to those who argue that the animal only mimics our language without understanding its meaning.

It is of course true that myriad questions about the mysterious horned rabbit remain unanswered—questions about its social biology, predation, seasonal migration, distribution, population dynamics, and the like. For example, experts will tell you that jackalopes have keen eyesight, acute hearing, an excellent sense of smell, and an uncanny ability to evade capture. But is it true, as Bill Alexander claims, that the jackalope "can also read minds, and some experts agree that they possess a certain ESP capability"? Is Bruce Larkin correct when he asserts, in his pamphlet *50 Facts about Jackalopes*, that jackalope saliva may be used to create waterproof ink, or that jackalopes sometimes snore loudly, or that "most serious Yeti trackers prefer to use specially trained jackalopes rather than bloodhounds"? Clearly, a great deal of further field research is necessary before we can develop a complete understanding of the elusive jackalope.

Despite many lingering questions, the impressive welter of details storytellers have offered about the horned rabbit makes perfectly clear that the jackalope exists, even if its numbers have been severely depleted since the nineteenth century when enormous herds of jackalopes roamed the Great Plains. Having researched the question for many years, I vehemently disagree with those who maintain that the jackalope, though now a staple of American folklore, has been driven to extinction. In "Cenotaph of the

Jackalope," Hermes Trismegistus "Hermester" Barrington helps us to think through this vexing question. "Its existence, while improbable, is not impossible, but it follows that the extinction of such a creature is likewise unprovable, unverifiable, unfalsifiable." This is the same impenetrable logic that has kept Bigfoot and Nessie alive for so long: they must exist because there is no definitive proof that they do not. And that sort of existence is the native province of tall-tale tellers, whose colorful yarns will keep the horned rabbit alive just as long as their imaginations hold out.

Chapter 3
The Classic Hoax

Hoaxes, to a greater or lesser degree, rely on our willingness, even our desire, to believe in the unusual occurrence, the remarkable coincidence, the unexpected result, the surprising fact, or the supernatural event. Audiences take pleasure in accepting the improbable as true—that is, in being deceived—since they find the optimism that produces such a belief, and the hope or faith that sustains it, preferable to the cynic's view that "nothing is new under the sun" and that anything exceptional, wonderful, or astonishing must be a trick.

—James Fredal, "The Perennial Pleasures of the Hoax"

During the summer of 1992, as I toiled endlessly at my dissertation in the sweltering heat inside a green board shack on a brambly mountainside on the outskirts of Charlottesville, Virginia, the surprising announcement was made about the discovery of a strange creature in the state next door. On June 23, the cover of the *Weekly World News* boldly proclaimed, "Bat Boy Found in West Virginia Cave!" The tabloid documented the story using captivating details, copious "evidence," and interviews with the scientists who had made the astounding discovery, but what I found irresistible was the stunning image of the bat-human hybrid. Bat Boy had giant amber eyes with wildly dilated pupils, separated by the tiniest of noses. Mouth agape midshriek, his teeth were tapered spikes, and his giant ears, completely out of proportion with his small face, rose to perfect points along the sides of his head. He was terrifying, but also somehow innocent, like a cute kid in hysterics after having fallen off a playground swing. And he was simultaneously hilarious. In an era before Photoshop and deepfakes became staples of our daily media consumption, Bat Boy was iconic. For the next fifteen years, until the paper edition of the tabloid folded in 2007, I delighted in wrapping every holiday and birthday gift in the alien abduction cover-ups and first-ever photographs of heaven regularly supplied by the *Weekly World News*.

As is the case with many media hoaxes, the trajectory of Bat Boy's adventures closely tracked the obsessions and anxieties of popular culture and contemporary politics. In November 2001, the paper announced that Bat Boy had met with President George W. Bush at Camp David, and soon after he stormed Afghanistan alongside the US Marines. A hero, you might think, until the December issue broke the disconcerting news that he had bitten Santa Claus. Eventually there came a torrent of wild narratives: Bat Boy being impersonated, hunted, even cloned. He had an affair with Jennifer Lopez. He flew the space shuttle. He guided US troops to the secret hideout of Saddam Hussein. His skillset was impressively versatile! In 2009 he received the prestigious MacArthur "Genius Grant" while simultaneously directing US Cyber Command and mentoring Tiger Woods through a rough patch in his golf game. While popular on the standup comedy circuit, Bat Boy was also chosen as the next pope. In 2019, he declared his candidacy for the US Presidency. Taylor Swift, who would soon be revealed to be carrying Bat Boy's child (the *Weekly World News* published the ultrasound as proof), hosted a fundraiser for his campaign. As of the summer of 2021, Bat Boy was on a world tour with his All Stars band, which features Bigfoot on drums (of course) and Manigator on the Hammond B3 organ.

As these delightful examples suggest, Bat Boy wasn't much of a hoax—if by hoax we mean a deception compelling enough to fool a great many people. But his story nevertheless proceeded according to the conventions of most media hoaxes. First of all, the *Weekly World News* wasn't much less reliable than a number of other tabloids of wide circulation, and in that sense it spoofed a sensational popular press that could hardly be relied on to tell the truth—and thus satirized anyone gullible enough to believe what they read in the tabloids. The Bat Boy narratives parodied familiar journalistic conventions, including the unquestioned objectivity of the reporter, and doctrinal reliance on the accounts of reputed first-hand witnesses. The stories also strategically appropriated and deployed the discourse of science, referring any doubts we might have about the bat-human hybrid to the experts who, after all, are better qualified than we to judge such things.

And then there was the uncanny way Bat Boy appealed to our imaginative fascination with hybridity and our unspoken hope that, even in an age of catastrophic biodiversity loss, some miracle of nature might yet exist to be discovered. If Bat Boy's antics became increasingly absurd, that only reflected the distressing reality that plenty of other things around us were also absurd. The fact that Bat Boy was phony bolstered his power to satirically expose so much else that was phony in American popular, political, and corporate cultures. Despite the troubling detail that he bit Santa, Bat Boy was wholly redeemed by his entertainment value. Like the jackalope, another hybrid creature conceived as a hoax but now enjoyed primarily by those who know better, Bat Boy made us smile, especially when the real world seemed even more alien and distressing than the notion that a bat-child hybrid might have emerged from a West Virginia cave.

A complex phenomenon, the hoax is related to other forms of dissembling such as the con, scheme, swindle, scam, sham, fabrication, flimflam, dodge, put-on, lie, grift, prank, sell, ruse, trick, rip-off, forgery, fraud, stunt, sting, and humbug. It is linked to mimicry, subterfuge, counterfeiting, plagiarism, deception, duping, diddling, charlatanism, bamboozling, hornswoggling, hoodwinking, and fabulation. To employ more recent (and less pejorative)

offshoots of the concept, the hoax is also akin to *détournement*, simulacrum, parafiction, culture jamming, and reality hacking—all of which include an essential element of political resistance or social critique. But why are we perennially fascinated by that which is (or, at least, *might be*) bogus? And how, perversely, does our fascination with distinguishing the authentic from the fake actually work to enable the hoax, an artful form of deceit that has existed for millennia? The hoax dates at least to Herodotus, who in *The Histories* (430 BCE) wrote of ants the size of foxes and Cyclopes that stockpile gold. In "Raising the Wind; Or, Diddling Considered as One of the Exact Sciences," so perceptive a student of human nature as Edgar Allan Poe (himself an inspired literary hoaxer) observed that "a crow thieves; a fox cheats; a weasel outwits; a man diddles. To diddle is his destiny. 'Man was made to mourn,' says the poet. But not so:—he was made to diddle. This is his aim—his object—his *end*." Poe considered the desire to hoax so irresistible as to be a defining characteristic of the human species.

In the previous chapter we had fun with the tall tale, a beloved American narrative genre that has helped to amplify and disseminate the legend of the jackalope. But at the headwaters of those braided, meandering stories is the thing itself: the mounted head of a rabbit with antlers, staring down from the wall above the bar, silently demanding from us an assessment of its authenticity. And while the jackalope's wide distribution in popular culture ensures that most people now see the joke coming, even today plenty of credulous folks are taken in by the horned rabbit; at the very least, they pause to wonder if maybe—just *maybe*—that strange and wonderful hybrid bunny really does exist.

When we hear the word *hoax*, we tend to think of it pejoratively, as a deliberate form of carefully orchestrated deceit that deprives its victim of something—whether it be their money, status, or self-respect. But the *Oxford English Dictionary* offers a more innocent gloss, explaining that to hoax is "to deceive or take in by inducing to believe an amusing or mischievous fabrication or fiction; to play upon the credulity of." While the hoax can occasionally be malicious, it is more often a form of play that seeks to generate amusement. The etymology of *hoax* remains obscure, but is likely an alteration of *hocus*, which in the seventeenth century referred to a conjurer or impostor. (The word may also be related to *hocus pocus*, a phrase documented from the 1640s.) In one way or another, the hoax makes a fool of anyone gullible enough to "fall for it"—an idiom that suggests the leveling power of the hoax. *If you think you're so smart*, the hoax seems to taunt, *then stake your reputation on whether or not I'm genuine*. Furthermore, a hoax usually has a prominent public dimension. It is staged with the specific intent of drawing attention to itself—attention that often focuses strategically on its own disputed legitimacy. Importantly, a hoax is not a scam, because it typically exists to amuse rather than defraud. A hoax isn't just a trick, but also a performance; and, like other kinds of performances, it is usually calculated to entertain.

Like the wobbling Jenga block tower that is the tall tale, the hoax ultimately exists to be discovered. While a forger or counterfeiter may wish to evade detection indefinitely, a hoaxer knows that exposure of the victim's naiveté can be realized only when the hoax is revealed. As my friend Lynda Olman writes in her book *Sins against Science*, a rhetorical study of scientific media hoaxes, "the defining feature of a hoax is the moment of embarrassment. In this moment the hoax reveals its devices, which amount to the reader's own assumptions, which the hoax has exploited to achieve its humiliating effect." The hoax is a peculiar form of deceit, in that it must fail in order to succeed. As Kevin Young observes in his outstanding book *Bunk*, "while the long con, when done right, need never have to end, the hoax is all about its discovery, hinted at all along. Hindsight is the hoax's best light. . . . This is why we are not just fooled, but made fools of, by the hoax—indicted by its revelations, not of what's true but of what we truly believe." Like the tall tale, the hoax can't get along without us, because it is our own unexamined preconceptions, our desperate desire to know the truth, our vain need to perform our discernment, that enable its success. Young is right that "we collaborate with the hoax, and collude with it; the hoaxer just gets there first, making unwilling coconspirators of us all." It is our desire to believe that there *might* be such a thing as a jackalope that breathes life into the jackalope hoax.

Its relationship to the tall tale also highlights several underappreciated aspects of the hoax. First, a hoax is usually artful; its success often requires impressive creativity or skill. For example, the paintings of the great art forger Elmyr de Hory—featured in Orson Welles's fascinating 1973 film *F is for Fake*—are certainly hoaxes, but there is a clear sense in which they are also artful, since their creation requires spectacular virtuosity, if not spectacular originality. De Hory went so far as to question the distinction between real and fake art, asserting in Welles's film that "if my work hangs in a museum long enough, it becomes real." If renowned art dealers and museum curators can't tell the difference between his work and a Picasso, he seems to ask, then how meaningful can that difference really be? (In a twist that de Hory would have enjoyed, he became so famous as an art forger that, after his death in 1976, fake de Hory paintings executed by other art criminals began to appear in the marketplace.) As Chris Fleming and John O'Carroll put it in "The Art of the Hoax," hoaxes "all involve some kind of artful deception, an aesthetically sophisticated act of trickery, of mimetic *artistry*. A good hoaxer is a very skilled reader and manipulator of textual genres and often specialized discourses."

The genre being manipulated in the case of the jackalope is taxidermy; the artful deception is creating a mount that appears to be a real animal. I have seen mounts so poorly executed that the ruse was immediately apparent, and others so expertly fashioned that the imaginary creature was made real through artifice alone. Like an art forger or counterfeiter, the taxidermist who fabricates a jackalope mount aspires to make their work indistinguishable from that of the taxidermist who produces, say, a well-crafted shoulder mount of a white-tailed deer. The tools and techniques of the trade are identical, but while one artist preserves a real animal, the other employs the conventions of taxidermy to create something fantastic. And isn't the production of the fantastic as much the goal of art as the representation of the real?

Again defying a common misconception of the hoax as necessarily malicious, it is vitally important that a hoax brings pleasure. Indeed, the understanding of hoax as mere swindle fails to appreciate the human psychology that is its primary engine. The long history of hoaxing proves that we will often pay to see something that might be a hoax when we would not pay to see something we know to be genuine, and that is because our desire to judge authenticity is often more engaging than authenticity itself. While I would enjoy standing before a painting by Matisse, Modigliani, or Renoir, what I would enjoy even more is standing before a painting that appears to be a Matisse, Modigliani, or Renoir and be asked whether I think the painting is genuine or a forgery painted by Elmyr de Hory. Perhaps perversely, the mere fact that the painting *might* be counterfeit—and that I am being asked to publicly render a judgment as to its authenticity—is more exciting to me than looking at an authenticated work by Matisse. In fact, if I am later informed that the painting is genuine, I may even find myself disappointed. A Matisse, however great, can be nothing more than it appears to be; a Matisse painted by a daring forger and sold for a great sum of money to art world elites who were utterly fooled by it, now *that* is something worth seeing! Likewise the jackalope. The mounted head of a deer soon ceases to be interesting because, however well-executed, it is nothing more than it appears to be. But an antlered rabbit? That is a *real* fake, one worth seeing not in spite of its artifice, but because of it.

Part of the pleasure we take in the hoax, then, is the exhilaration we feel in wondering whether what we see (or read, or hear) is genuine. As James Fredal observes, "A hoax that deceives no one fails. But to an equal degree, a hoax that is never discovered also fails. Its truth wants out. A hoax must therefore balance probabilities and improbabilities, presenting itself to its audience as both real and fake, such that the audience is required in the process of interpreting the event, to determine which of these two readings is the correct one and preferably accepting each in turn." When we observe the Matisse we are passive observers; when we observe the Matisse that might be a de Hory forgery, we are active. This activation of attention is a key dynamic of the hoax's allure.

LEFT: Douglas and Ralph Herrick, of Douglas, Wyoming, are reputed to have been the first taxidermists to make a jackalope mount during the 1930s. This excellent full-body mount was the last jackalope ever fabricated by Ralph Herrick. Circa 2004. Courtesy of Wyoming Pioneer Memorial Museum, a division of Wyoming State Parks and Cultural Resources. *Photo by the author.*

RIGHT: The charming old Hotel LaBonte, in Douglas, Wyoming, was built in 1914. The first jackalope mount ever made was displayed on the wall of the hotel's bar for decades, before it vanished in the late seventies. *Photo by the author.*

ABOVE: Mermaids are among the oldest taxidermy hoaxes. The most famous and profitable of these, the Feejee Mermaid, was displayed by P. T. Barnum in his American Museum in New York City. Gift of the Heirs of David Kimball, 1897. *Image © President and Fellows of Harvard College, Peabody Museum of Archaeology and Ethnology, 97-39-70/72853.*

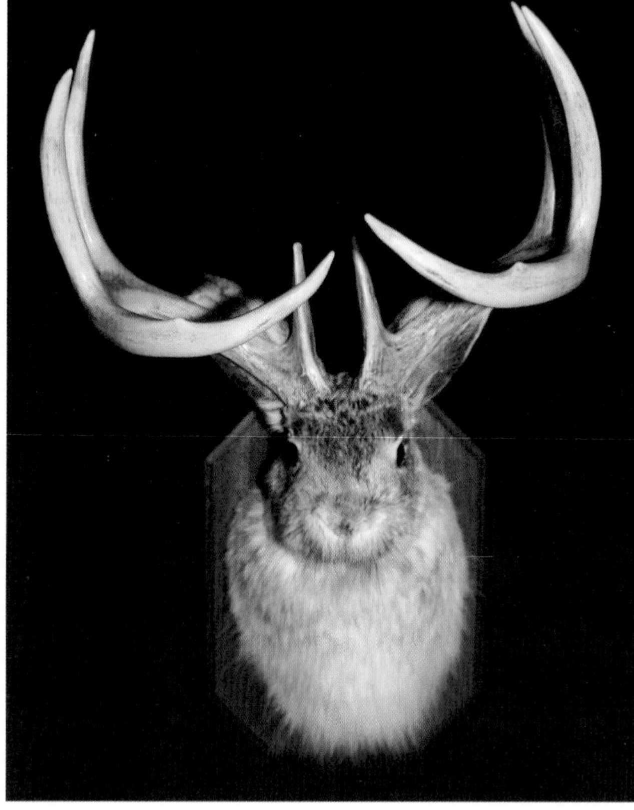

LEFT: Frank and Dianne English, of Rapid City, South Dakota, are among the most prolific of jackalope makers. This "World Record Jackalope" (a fourteen-point buck!) is one of the many custom jackalope mounts fabricated by Frank English in his home workshop. *Photo by the author.*

LEFT: Postcards are among the oldest and most popular items of jackalope memorabilia. This vintage card, circa late 1940s, came from the studios of photographer Harold Sanborn. It was immensely popular, with at least 17,000 copies made over the life of the negative. *Courtesy of John Meissner, Sanborn Research Centre and Estes Park Archives. Photo by Kyle Weerheim.*

BELOW: The tradition of the jackalope postcard is kept alive today by contemporary artists including Chet Phillips, whose delightful "Greetings from Austin" card collection features a playful jackalope who always manages to have a good time. *Courtesy of Chet Phillips.*

ABOVE: A noted roadside attraction and undoubtedly the world's most famous emporium for jackalope kitsch, South Dakota's Wall Drug sells jackalope T-shirts, key chains, postcards, shot glasses, and shoulder mounts by the dozen. *Courtesy of Sarah Hustead. Photo by the author.*

BELOW: President Ronald Reagan displayed a buck-and-doe jackalope shoulder mount in his Rancho Del Cielo home in Goleta, California. In this photo, taken on November 25, 1987, Reagan tries to pull one over on National Security Advisor Colin Powell, whose laughter shows that he knows better. *Courtesy of the Ronald Reagan Library, image C43826-25.*

RIGHT: Entitled *Jack*, this imaginative assemblage piece by Denver artist Bill Nelson also includes a jackalope mount made by Michael Herrick (the son of the jackalope's inventor) and floral work by Emily Marchalonis. *Courtesy of Bill Nelson.*

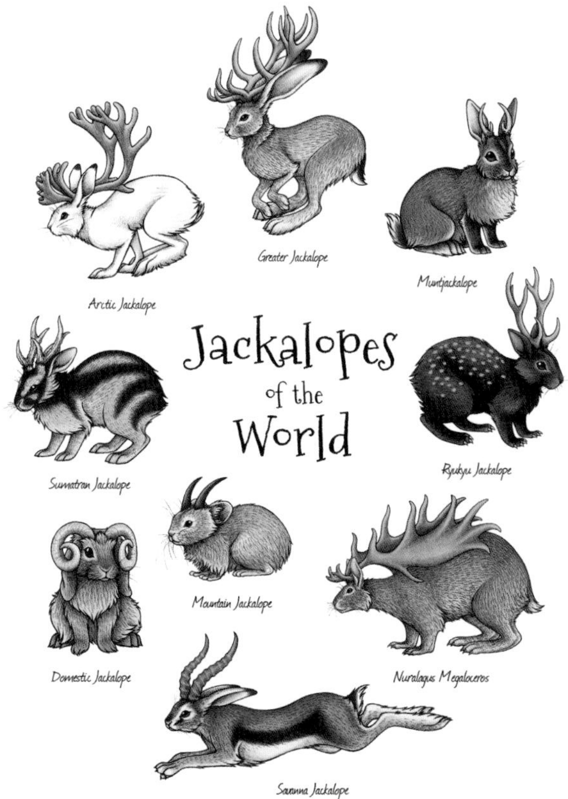

LEFT: In *Jackalopes of the World*, British artist Lyndsey Green experiments with how horned rabbits "might have evolved and adapted to live in different climates." *From* Jackalopia: The Pocket Guide to Jackalopes of the World *(2016). Courtesy of Lyndsey Green.*

LEFT: Andrew Williams's *Lope and Change* jackalope campaign image riffs on the iconic *Hope* image created by artist Shepard Fairey during Barack Obama's 2008 presidential run. *Courtesy of Andrew Williams. http://jackalope.xyz/.*

BELOW: Hannah Yata's 22" x 28" oil on canvas painting, *Dahlia* (2018), is characteristic of how her surreal, psychedelic work creates dreamscapes. The piece shows that the jackalope can morph into the fantastical when reimagined by contemporary artists. *Courtesy of Hannah Yata.*

LEFT: Jackalopes of every imaginable style and design may be seen in body art. This jackalope tattoo is by artist Summer Orr, of Reno, Nevada. *Courtesy of Summer Orr.*

BELOW: Nevada artist Reena Spansail's remarkable 5" x 7" watercolor *Faster Than the Speed of Light* (2017) depicts a cosmic jackalope in full flight through a sidereal field of spinning stars. *Courtesy of Reena Spansail.*

Toronto-based artist Kari Serrao creates encaustic paintings, which are made using heated, pigmented beeswax. This large, 36" x 48" piece, *Morning Music*, captures the whimsy of the jackalope, while also putting the horned rabbit into relationship with other woodland creatures. *Courtesy of Kari Serrao.*

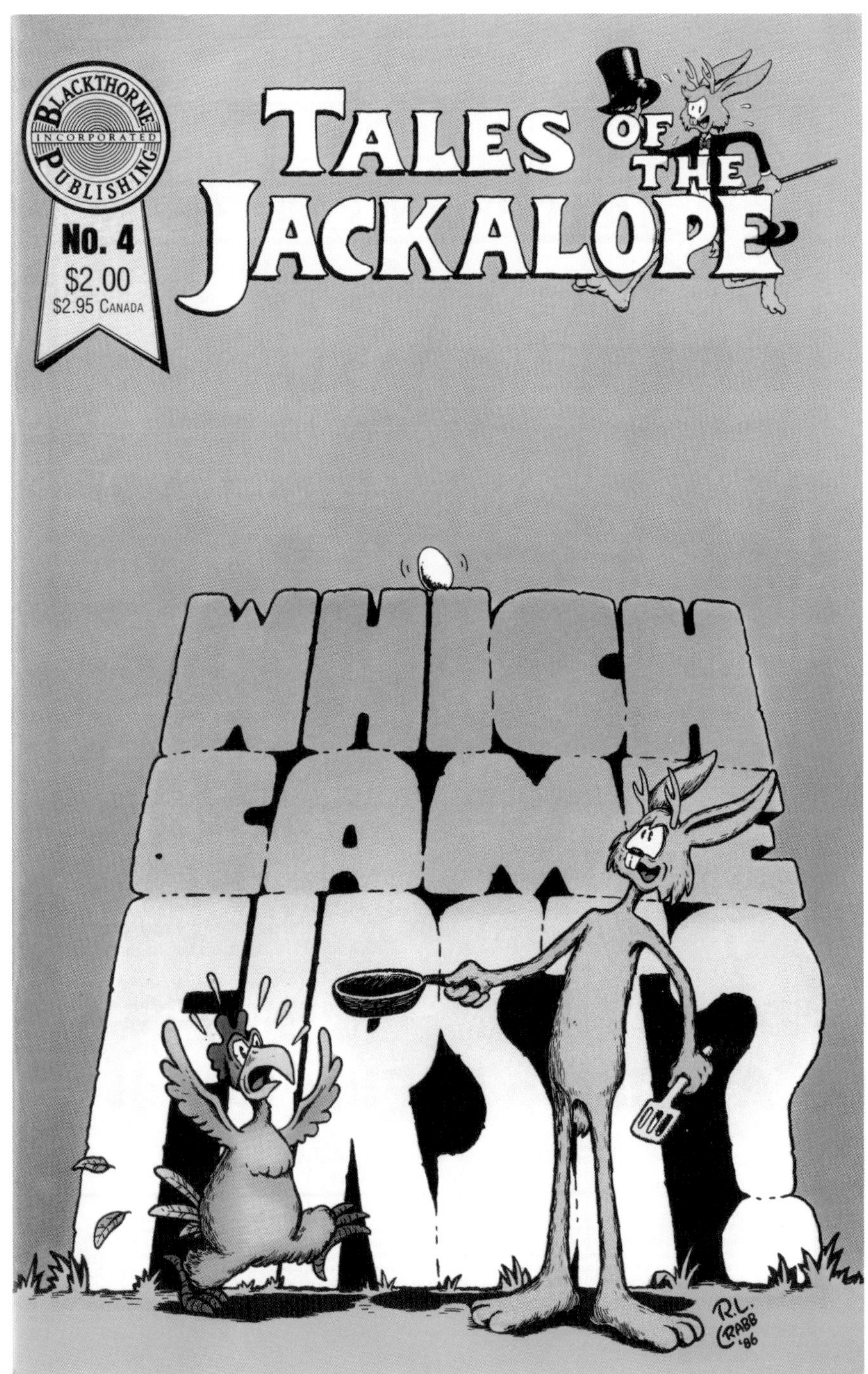

During the 1980s, northern California cartoonist R. L. Crabb created Junior Jackalope, a sly, politically progressive character who was the hero of the *Junior Jackalope* and *Tales of the Jackalope* comic book series. Cover of *Tales of the Jackalope* #4 (1986), Blackthorne Publishing. *Courtesy of R. L. Crabb. Photo by Kyle Weerheim.*

Seattle-based illustrator Lily Seika Jones combines the strangeness of the jackalope with the precision of natural history illustration in this anatomical depiction of the horned rabbit. *From* The Compendium of Magical Beasts, *by Veronica Wigberht-Blackwater, © 2018. Reprinted by permission of Running Press Adult, an imprint of Hachette Book Group, Inc.*

LEFT: This creepy specimen, *Lepus temperamentalus* (Jackalope), is on display as part of London-based artist Alex CF's grotesque and beautiful Merrylin Cryptid Museum. Image © Alex CF 2014. *Courtesy of Alex CF.*

BELOW: Prominent naturalist Ernest Thompson Seton was well aware of horned cottontails, and drew a number of specimens collected in 1916 and 1917, long before the first jackalope hoax mount was made by Douglas and Ralph Herrick. *From Seton's* Lives of Game Animals, *vol. 4, pt. 2 (New York: Doubleday, Doran & Company, 1929), 811. © Julie A. Seton, E. Micah Barber, and Sheryl W. Barber. Photo by Kyle Weerheim.*

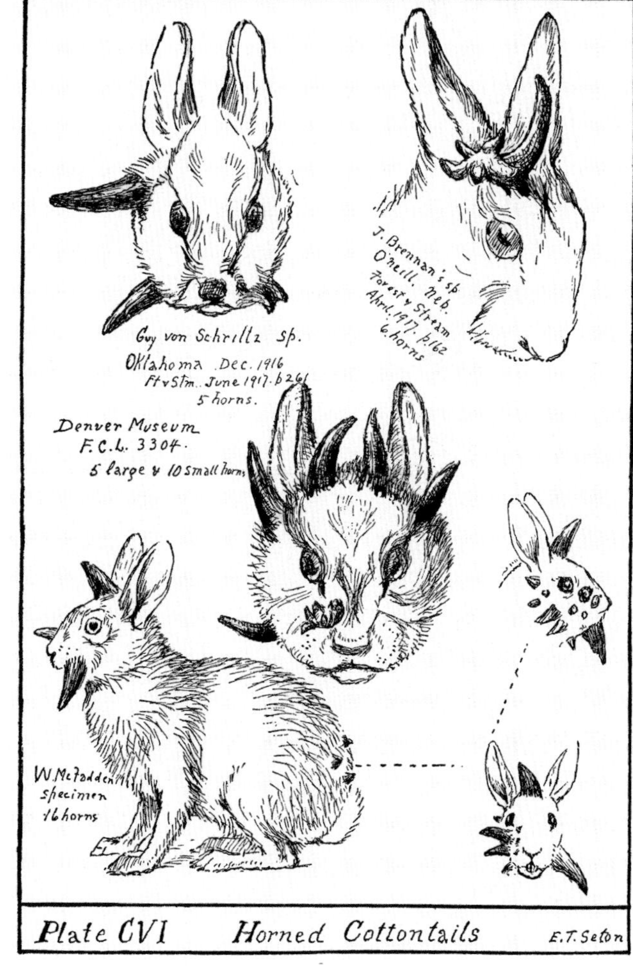

RIGHT: Among the most ancient of jackalope precursors is Al-Mi'raj, described by thirteenth-century Persian naturalist Abu Yahya Zakariya' ibn Muhammad al-Qazwini in his geographical encyclopedia, *'Ajā'ib al-makhlūqāt wa gharā'ib al-mawjūdāt (Marvels of Things Created and Miraculous Aspects of Things Existing).* Originally published circa 1280. *From the updated manuscript in the Bavarian State Library, published 1750–70. Image from Wikimedia Commons.*

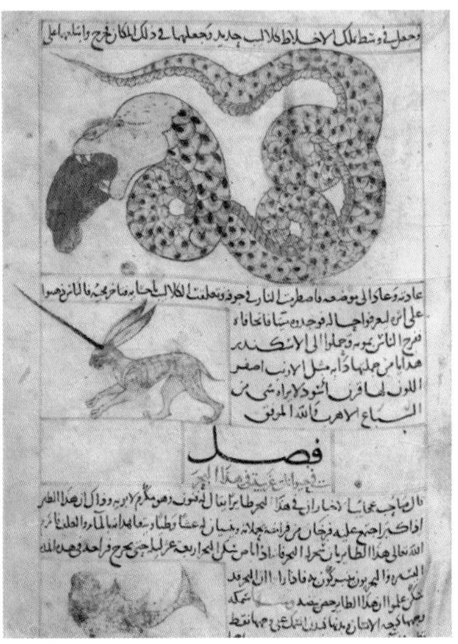

ABOVE: European Renaissance naturalists depicted *Lepus cornutus*, the horned hare, which they believed to be a distinct species. Flemish manuscript illuminator Joris Hoefnagel illustrated this horned rabbit situated between a normal rabbit (on the right) and a rabbit patterned after Albrecht Dürer's 1502 watercolor *Young Hare* (on the left). Plate 47, *Animalia Quadrupedia et Reptilia* (*Terra*), circa 1575–1580. *Courtesy of the National Gallery of Art. Gift of Miss Lessing J. Rosenwald, accession number 1987.20.6.48.*

Seventeenth-century Flemish painters Jan Brueghel the Elder and Peter Paul Rubens collaborated on the exquisite painting *The Virgin and Child in a Painting surrounded by Fruit and Flowers*, for which Brueghel painted the settings, garlands, and animals, while Rubens created the human and angelic figures. Note the horned rabbit depicted in the foreground right. Circa 1617–20. Oil painting with panel support. *Image courtesy of the Museo del Prado, Madrid, Spain. Inventory no. P001418.*

ABOVE: As late as the nineteenth century, the horned rabbit remained a source of fascination to European naturalists. The realistic, irregularly branching growths on French engraver Robert Bénard's *Le Lièvre Cornu* suggests that the model for his illustration may have been a rabbit stricken with Shope papillomavirus. Plate 61, *Mammalogie, ou, Desctription des Espèces de Mammifères*, by Anselme-Gaetan Desmarest and illustrated by Robert Bénard. 1820-22. *Courtesy of the Biodiversity Heritage Library. Image from flickr.*

LEFT: Dr. Richard E. Shope pioneered the study of "horned" rabbits. This photo of Shope with his graffitied jalopy, "Galloping Asthma," was taken as Shope and his wife, Helen, traveled from their home in Iowa to the Rockefeller Institute, in Princeton, New Jersey, where Shope would have a distinguished career as a virologist. He labelled the photo "Tire Trouble South of Tipton, Iowa." August 21, 1925. *Courtesy of Nancy Shope FitzGerrell.*

An adult male cottontail (*Sylvilagus floridanus mearnsii*) from Topeka, Shawnee County, Kansas, with growths caused by an extreme case of Shope papillomavirus, from 1989. Specimen on deposit at the University of Kansas Biodiversity Institute, Lawrence, KS. *Courtesy of Heather A. York.*

After many years of following the trail of the jackalope, the author at last takes on the challenge of making a jackalope himself. The imperfect result is "Paxton," who was fabricated by the author at a jackalope making workshop in the Mission District of San Francisco on January 11, 2020. *Photo by Kyle Weerheim.*

In addition to the pleasure of being engaged in this absorbing act of discernment, there is also pleasure in the eventual revelation of the hoax. "Holy shit, are you kidding me? That Matisse is really a fake?" You can easily imagine the viewer's delight! And that wonderful moment of revelation possesses a double charm. First, there is the joy we experience in surprise, in ascertaining the truth about the counterfeit. But there is also a subsequent pleasure in admiring how de Hory's consummate skill allowed him to succeed in creating something so beautiful, exact, and convincing.

Finally, the hoax brings us a new kind of pleasure once we know its secret but others do not. Even if I am aware that the would-be Matisse is counterfeit, I will enjoy watching you try to figure it out for yourself. And in this process there is no outcome that will not be good fun for me. If you speculate correctly that the painting is fake, I will relish seeing you experience the same pleasurable process of discernment and revelation that I did. If, instead, you are fooled by the forgery, I will enjoy myself even more. Better still if you hold forth about Matisse's distinctive use of vibrant color, the unmistakable speed and confidence of his brushstrokes, his inimitable mastery of the alla prima technique. The longer you pontificate, the sweeter will be the moment when the hoax is exposed, and you with it. Your embarrassment will be as wonderful for me as it is awful for you. Please don't feel badly. After all, some other sucker paid big bucks for this painting before it was discovered to be counterfeit. Your pride may be hurt, but the hoax has administered a dose of humility that may serve you well down the road. Do I feel guilty about having let you go on talking so long when I was fully aware that you had "fallen for it," when I knew that with every word you were climbing a ladder of certainty from which your inevitable fall would be that much more chastening? Yes, I do. Was it worth it? Absolutely.

The jackalope hoax mount is my Elmyr de Hory Matisse; the gallery in which it hangs is any bar or pool hall, restaurant or bus station, office or garage anywhere in the world. Those not already in on the joke will delight in seeing the antlered bunny for the first time; it is amusing to wonder if the thing is real, but whether it is real or not, it will be fun to find out the truth.

People take joy in this wondering, and another kind of delight in discovering the reality of the fake, which causes them to admire the imagination and craft necessary to "mount" the hoax in the first place. If the moment of revelation is humbling, that is simply the price of admission to the jackalope's secret. There will be ample recompense in the years ahead as the embarrassment of others brings unending pleasure. This is why so many people who are taken in by the jackalope mount soon move from being the hoax's victim to becoming its perpetrator by purchasing and displaying a hoax mount of their own.

Hoaxes take myriad forms, but among the most famous and fascinating hoaxes in history are science hoaxes. The Great Moon Hoax of 1835 was propelled by the *Sun*, one of the New York "penny papers" (precursors to tabloids like the *Weekly World News*), which published a series of articles describing in detail how a revolutionary new telescope had made possible the astounding discovery of life on the moon—life reported to include bipedal beavers, unicorns, horned bears, and, as the pièce de résistance, winged man-bats (ancestors of Bat Boy, no doubt). Entirely by design, heated public arguments regarding the veracity of the newspaper's claims caused its circulation to skyrocket.

The Cardiff Giant hoax, perpetrated in 1869, involved the remarkable discovery of a ten-foot-tall "petrified man" on a farm in Cardiff, New York. The giant was not an ancient artifact but rather the work of tobacconist George Hull, who had gone to considerable expense to have the figure fashioned, transported, and buried in anticipation of its "discovery." Controversy over the artifact's authenticity was widely publicized, and the crowds that flocked to the farm to see the marvel were charged handsomely for the privilege. So sensational was the find that the greatest hoaxer of the nineteenth century, P. T. Barnum, offered to buy the figure for $50,000. When he was refused, Barnum, ever the consummate entrepreneur, simply commissioned his own fake Cardiff Giant, which he then displayed while accusing the original Cardiff Giant of being inauthentic. The men to whom Hull had sold the original giant sued Barnum but the courts, in a moment of delicious irony, ruled that Barnum, the great faker, could not be held legally liable for calling a fake a "fake."

Among the most famous paleoanthropological hoaxes was Piltdown Man. In 1912, amateur anthropologist Charles Dawson claimed to have discovered a fossil that was the long-sought "missing link" between man and ape. The find stirred excitement and debate in the scientific community, dividing expert opinion regarding the fossil's legitimacy. The revelation of this hoax would have to wait until 1953, when a team of scientists proved conclusively that the fraudulent fossil was a composite of the skull of a human, the jaw of an orangutan, and the fossilized teeth of a chimpanzee.

Even the much-studied Sokal affair, which occurred in 1996, qualifies as a science hoax. Physicist Alan Sokal submitted an article entitled "Transgressing the Boundaries: Towards a Transformative Hermeneutics of Quantum Gravity" to the prominent cultural studies journal *Social Text*. In it, Sokal asserted a great deal of complete nonsense, including that quantum gravity may be understood as merely a linguistic and social construct. The unsuspecting editors of the journal published Sokal's article, after which ensued a robust debate regarding a range of issues, including whether scholars in the humanities too often appropriate scientific concepts without fully understanding them.

It is little wonder that so many hoaxes co-opt science, as science is a realm of substantial cultural authority characterized by specialized discourses not easily accessible to a lay audience. We trust scientists to validate many things we will never be able to judge for ourselves, and so an adroit hoaxer will often deploy the rhetoric of science to execute their ruse. The Great Moon Hoax, for example, depended for its credibility on the newspaper's claim that John Herschel, who was among the most prominent astronomers of his day, had been the discoverer of the amazing forms of lunar life. The headline in the *Sun* read "GREAT ASTRONOMICAL DISCOVERIES LATELY MADE BY SIR JOHN HERSCHEL, L.L.D. F.R.S. &C." Herschel, who knew nothing of the hoax, was conveniently incommunicado when the story broke, off making actual astronomical observations in South Africa. Although the Cardiff Giant hoax was perpetrated by a nonscientist, the notoriety of the find was fueled when scientists were invited to examine the would-be artifact and render their expert judgment. While the perpetrator of the Piltdown Man hoax remains uncertain (whether Charles Dawson himself, or one or more of a number of other suspects), this hoax capitalized on the tremendous hunger among scientists to prove the hypothesized "missing link" between ape and human. Professor Sokal's hoax is perhaps the clearest example of how the discourse of science may be deployed to create a culturally authoritative narrative that can disorient a nonspecialist audience. Sokal's scientific expertise was undeniably impressive, and nothing greases the wheels of a hoax like expertise. This is the case whether the scientific expert perpetrates the hoax, as did Sokal, or finds themselves its unwitting accomplice, as did Herschel.

Ironically, even when the consensus reached by scientific experts who examine a would-be hoax is that the phenomenon is indeed a fraud (as was the case with the Cardiff Giant), the superficial legitimacy gained through the involvement of reputable scientists can still increase the prominence of the hoax and, in many cases, its profitability as well. So powerful and disarming is the discourse of science that it has even been used to make a hoax of that which is not a hoax. Among my favorite examples of this is the alarmed response provoked by the breaking news (and it breaks every now and then, as this nonhoax hoax is periodically revived) that dihydrogen monoxide has been found in the city's pipes, or that it has been discovered to be the main component of acid rain, or that in its heated state it can cause severe burns. This would all be shocking, if dihydrogen monoxide were not simply H_2O. By employing scientific nomenclature to obscure the fact that they are describing nothing more than water, the perpetrators of this parody succeed in making us laugh while also satirizing our troubling lack of scientific literacy.

A subset of science hoaxes involves mysterious specimens of one kind or another, and on many occasions these "specimens" have been produced by wily taxidermists. As a devotee of the jackalope I am enamored of taxidermy hoaxes, which have a long history. And while many people are familiar with blatant taxidermy gags such as the fur-bearing trout (a hoax dating to the 1920s), or the much more recent assalope (the rump of a deer mounted so as to resemble a face), taxidermy hoaxes are surprisingly widespread and often skillfully executed.

A little-known taxidermy hoax that I admire both for its playfulness and its intelligence concerns that rarest of birds, the Bare-fronted Hoodwink. Matthew Meiklejohn (1913–1974) was a professor of languages at the University of Glasgow, Scotland, and also an amateur ornithologist of repute. In 1950, Professor Meiklejohn published an article in *Bird Notes* (the journal of the Royal Society for the Protection of Birds) about an extremely uncommon avian species. "I am impressed by the number of records of birds partially seen or indeterminately heard," he begins, "and it seems evident that the majority of these records are attributable to a single species—the Hoodwink—which I propose to name *Dissimulatrix spuria*." Meiklejohn goes on to offer the cheeky explanation that this elusive species is "generally recognisable by *blurred appearance* and extremely rapid flight away from the observer." The Hoodwink's defining quality is that it has never been fully observed. It has, however, been photographed: "It is the brown blur that passes rapidly from right to left in all ornithological films." Meiklejohn's piece is carefully structured to resemble a proper ornithological article, including sections on habitat, voice, display, breeding, food, distribution, and migration. Within each section, however, fantastic stories are told. Where does the Hoodwink roost? "Among rows of bottles *in bittern-like posture*." The bird's call? "In many records of bird song, also, the Hoodwink can be heard in the background, imitating the barking of dogs, the hooting of automobiles, the pleasant drone of the farm tractor, etc." Its breeding habits? "Details little known, but undoubtedly sometimes reproduces by *binary fission*; many reliable ornithologists, watching the Hoodwink, have seen it split into two halves and fly away in opposite directions."

Meiklejohn's imaginary bird was aptly named. In its figurative usage, the word *hoodwink*, which dates to the early seventeenth century, is defined by the *Oxford English Dictionary* as "To blindfold mentally; to prevent (any one) from seeing the truth or fact; to 'throw dust in the eyes' of, deceive, humbug." Better still was the scientific name he chose for this bird so rare as to not quite exist. The genus name *Dissumlatrix* of course suggests *dissimulation*, while the species name *spuria* invokes the *spurious*—that which appears to be but is not actually valid. And the lie of this hoax reveals a deeper truth, for we often catch only a glimpse of a bird in the bush or in rapid flight, a bird so mysterious that our repeated near-misses at positive identification over time suggest the idea that the bird simultaneously does and does not exist.

It is this latter fact that the Royal Scottish Museum (now part of the National Museum of Scotland) celebrated when, on April Fools' Day, 1975—just a few months after Meiklejohn's death—the museum mounted an exhibit dedicated to the Hoodwink. The display included blurry photos of birds in flight away from the viewer, as well as birds frustratingly unidentifiable because they were only partially in frame. Bob McGowan, the museum's Senior Curator of Birds, informed me that one of the exhibit's photographs captured "a group of ornithologists 'just missing sight of the bird.'" To make the real exhibit celebrating the fake bird more believable, a Hoodwink was actually fabricated by a professional taxidermist. According to McGowan, "this specimen comprised parts of other small, brown nondescript birds." When I contacted the museum to ask if this fanciful piece of taxidermy might still be hidden away in their archives seventy years after its display, I received an enthusiastic reply from curator Sankurie Pye, whose brilliant response demonstrated her appreciation for the hoax. "We still have the Hoodwink," she reported, "and, incidentally, also a spotted gruck!" You are forgiven if you find the gruck unfamiliar; like the hoodwink, it does not exist.

Among the most historically significant taxidermy hoaxes are mermaids, which have been fabricated, sold, and displayed for centuries. As Jan

Bondeson shows in his detailed study of the phenomenon, until well into the nineteenth century mermaids were not only widely believed to exist, but were frequently spotted by sailors (who may have been seeing dugongs or manatees) and even reported to have been captured by fishermen. A live mermaid was said to have been caught off Edam in Holland in 1403; another was taken in 1531 in the Baltic; at Ceylon in 1560; in 1718 off Amboine; in 1749 near Nyukoping, in Jutland; around 1800 at the Isle of Man; and in 1809 off the Scottish coast, to name just a few. No less accomplished a naturalist than Linnaeus was fascinated by the would-be species, even mentioning a "siren" in the tenth edition of his monumental *Systema Naturae* (1758). Because widespread belief in mermaids was buoyed by sailors' stories, there existed a special opportunity for skilled taxidermists to supply material specimens to satisfy the public's imaginative appetite. Where credulity reigns, the hoaxer flourishes.

The most famous of these many counterfeit mermaids was purchased by sea captain Samuel Barrett Eades, who bought it from Dutch sailors, who were said to have purchased it from a Japanese fisherman oblivious to the natural wonder he had drawn up in his nets. (It is more likely that the Dutch sailors were drawn into the Japanese fisherman's net.) We do know that Eades paid the princely sum of 5,000 Spanish dollars for the three-foot-long object, which he brought to London and exhibited publicly in 1822. Although the mermaid became a sensation in London and toured the provinces the following year, even at a shilling a peek Eades would never recover the incredible sum he had paid for the specimen. (Not to mention that he raised the money to purchase the mermaid by selling, without permission of its owner, the ship under his charge!)

When Captain Eades died in the early 1840s, he left the mermaid to his son, who apparently wasted little time in selling the curiosity to Moses Kimball, owner of the Boston Museum. In 1842 Kimball took the specimen to New York to show it to his friend, the showman and self-declared "Prince of Humbug," P. T. Barnum. A deal was soon struck: Barnum would lease the mermaid for $12.50 per week. While evidence suggests that Eades believed his mermaid to be real (after all, he paid a fortune for it), the authenticity of the artifact was of no concern to Barnum. At the time Kimball approached him, Barnum had recently purchased Gardner Baker's American Museum in New York, which he ran as Barnum's American Museum from 1842 until it burned to the ground in 1865. One of the most popular attractions of its time (at the height of its popularity it hosted fifteen thousand visitors daily), the "museum" was an early example of infotainment, seamlessly blending bona fide natural history displays with sensational hoaxes and spicing it all with the bizarre sideshow fare that would lead Barnum to obscene wealth as a circus impresario. Under Barnum's direction, the American Museum featured taxidermy and osteology hoaxes, including an elephant skull billed as the cranium of a Cyclops, and a narwhal tusk mounted on a goat skeleton in order to create a unicorn. Among Barnum's later attractions was Jumbo, a huge African bush elephant that was a true crowd pleaser. After Jumbo was killed in an accidental collision with a railroad train in 1885, Barnum again turned to taxidermy, directing that Jumbo's massive skin be stuffed. Even in death Jumbo proved to be a pachyderm profit mill.

The glory of Barnum's main attraction, now rebranded the "Feejee Mermaid," was not that it was unique (remember, the very same specimen had already been exhibited in London) or that it was beautiful (despite Barnum circulating drawings depicting the mermaid as a voluptuous, bare-breasted woman with flowing locks, it was strikingly ugly). The specimen was only part of the ruse. As with the jackalope hoax, what mattered most was narrative, and storytelling calculated to satisfy an audience's thirst for the fantastic was Barnum's forte. First, he seeded various newspapers with fraudulent letters announcing that a famous British naturalist, Dr. Griffin (of the nonexistent London Lyceum of Natural History), would soon visit the US, bringing with him a sensational curiosity: a mermaid! It was reported to be unlikely that the good doctor could be prevailed upon to allow the public to witness this wonder of nature; just as Barnum predicted, the mermaid's inaccessibility produced a pent-up desire to see it. Next, Barnum enlisted his assistant, Levi Lyman, to impersonate Dr. Griffin. It was a role Lyman (could his name be more perfect?) played with panache, delivering many well-attended public lectures in which he expounded his scientific theories, establishing a fictitious legitimacy that ensured the success of his boss's hoax. In addition to casting Lyman in the role of lead scientist, Barnum supplied a raft of pseudoscientific documentation to support his mermaid ruse, including an elaborate narrative of the heretofore unknown creature's dramatic capture in the wild South Pacific. He exhibited the Feejee Mermaid for one week at the large Concert Hall in New York before moving it to his American Museum, where attendance soon tripled. As late as 1859, Barnum would take his mermaid with him on a tour of England—where, ironically, the same specimen had been displayed by Captain Eades back in 1822.

The now-legendary Feejee Mermaid was simply a well-executed taxidermy hoax consisting of the mummified torso and head of a juvenile monkey carefully sewn to the posterior end of a fish. It was a technique that had been used before by earlier taxidermists to produce other fraudulent mermaids. However well-executed Barnum's novel object might have been, its success depended on his uncanny ability to enfold the material object in layers of expertly crafted narrative—including, importantly, scientific narrative. The sensational success of the Feejee Mermaid helps explain how a horned rabbit invented in rural Wyoming managed to migrate not only across the American West, but around the world. A direct descendant of the Feejee Mermaid, the jackalope is a hybrid and a hoax, and one that has generated plenty of fabulous stories of its own.

While today we may wonder at the surprising credulity of those who "swallowed" hoaxes like the Feejee Mermaid, during the eighteenth and nineteenth centuries strange wonders of nature were reaching European and American naturalists from every corner of the globe. A case in point is the first platypus specimen, sent from Australia by Governor John Hunter to English zoologist George Shaw at the British Museum (where the specimen is still held today). Shaw struggled to accept that this strange animal could be real, writing in 1799 that "of all the Mammalia yet known [the platypus] seems the most extraordinary in its conformation; exhibiting the perfect resemblance of the beak of a Duck engrafted on the head of a quadruped. So accurate is the similitude that, at first view, it naturally excites the idea of some deceptive preparation by artificial means." The following year Shaw added, in the commentary on the platypus in his *General Zoology*, that "it was impossible not to entertain some distant doubts as to the genuine nature of the animal, and to surmise, that, though in appearance perfectly natural, there might still have been practised some art of deception in its structure." In other words, the platypus was so bizarre that Shaw speculated it might be a taxidermy hoax.

Perhaps we should give Shaw the benefit of his doubt. After all, the platypus is an animal that has a bill resembling a duck's, lays eggs like a bird, and nurses its young with milk, as does a mammal. The animal charms us with its strangeness, its apparent hybridity, its effortless resistance to any attempt to neatly taxonomize it. We live in a world in which the ancient, lobe-finned coelacanth, a fish thought to be extinct for sixty-six million years, can be caught by an unsuspecting fisherman off the coast of South Africa. A world where anadromous white sturgeon in the Columbia River grow to twenty feet in length, 1,500 pounds in weight, and may live a century. A world where the iridescent hummingbird hovering at your backyard feeder is less than four inches long, weighs .15 of an ounce, and beats its wings—which move in a figure-eight pattern to allow thrust on the upstroke as well as the downstroke—up to eighty times per second. In such a world, we might just as easily believe in a jackalope as a platypus. As both the fake Feejee Mermaid and the real platypus remind us, our credulity is tested not only by the hoax but also by the truth.

Andy Robbins's *Field Guide to the North American Jackalope* (2019) is among the most delightful of jackalope hoaxes. The thirty-two-page booklet is deliberately structured as a field guide, complete with coverage of jackalope taxonomy, species distribution, and seemingly-credible information about the animal's diet, social biology, habitat, and lifespan. Robbins also includes helpful sections on how to locate a jackalope, jackalopes in popular culture, and jackalope hunting and conservation. Because he is a talented visual artist, his *Field Guide* deftly models the work of a science illustrator, including a range of lovely line drawings of the jackalope, not to mention illustrations of its skull, tracks, scat, horn variations, as well as subspecies distribution maps.

Robbins skillfully deploys a staple technique of the tall tale-teller or hoaxer: he strategically draws our attention to the issue of credibility. Throughout the *Guide* he has inserted "Fact or Fiction?" boxes, within which various claims are evaluated for their accuracy and sometimes debunked. *Jackalopes can run 55 mph.* FACT. *Jackalopes are the product of an outlandish lab experiment.* FICTION. Robbins even knows the value of subtly hedging a judgment. *Jackalopes only mate during lightning storms.* "Legends say that jackalopes conceive solely during storms with significant atmospheric electricity," he explains. "While apocryphal, such accounts do contain a grain of truth. Jackalopes have shown a preference for evenings of heavy cloud cover or turbulent weather in which to conceal their mating rituals." PARTIAL FACT. By expertly mimicking the familiar generic conventions of an actual field guide, Robbins succeeds in contextualizing the fantastic (the jackalope as a species) within the familiar (the structure, language, and visual aesthetic of a field guide). From the first page of his *Guide* to the last, Robbins plays it straight, maintaining the deadpan delivery Mark Twain claimed was essential to any successful tall tale or hoax.

I tracked Andy Robbins down and asked if he would be willing to talk about how he crafted his *Field Guide*. My call to Andy found him at his home in rural Wyoming, where he does a great deal of painting that is surreal and abstract—an artistic proclivity that would seem to make him an unlikely candidate to execute the lovely whimsical line drawings which grace the *Field Guide to the North American Jackalope*.

"I got interested in the jackalope through taxidermy," Andy told me. "I had the opportunity to hang out with a friend of mine who's a skilled taxidermist, and I told him 'I want to make a jackalope.' And he gives me this sidelong look, like 'really?' But that's the kind of stuff that interests me. I thought, why are we doing four hundred whitetail shoulder mounts this year? Let's do some cool stuff!" I asked Andy what he found most compelling about the horned rabbit. "The jackalope is a simple visual gag. A punchline," he replied, explaining that, as an artist, he wanted to take "a simple visual joke and add details to sort of enlarge that world." When asked how he settled on the gentle, fanciful visual aesthetic of the *Guide*, Andy told me that "drawings make it seem more real than Photoshop." He chose to work in simple line drawings because inexpensive field guides are often in black-and-white.

I was especially interested in Andy's key choice to model his project after the genre of the field guide. As a visual artist, he said he liked the format because it allowed him to do lots of illustrations relative to the amount of text included. More important, he relished the feeling of realism provided by the genre. "I wanted parents and their kids to be able to go into a bookstore, and if the kids asked if jackalopes are real the parents could say, 'Of course they're real. If they weren't, why would there be a field guide?'" But Andy didn't want hyperbole to blow the cover provided by the nature guide format. "I wanted to use stuff that makes sense within the natural world," he said. In fact, Andy resisted when his publisher requested that he make his work more outrageous, maintaining instead that the humor should remain dry. He was inspired in part by Bigfoot field guides, which he said took themselves so seriously that he found them hilarious.

As fellow jackalope aficionados, Andy and I talked for a long time, and by the end of our conversation we had begun to wax philosophical. I asked if he thought the jackalope mount offered any social critique, or if it was just pure fun. "If the jackalope is a commentary on anything, it could be a commentary on why we desire trophies of animals that we've killed. The jackrabbit being an innocent little rabbit is not something to be proud of. So you're taking that kill and trying to make it something that looks more majestic." I agreed with Andy, adding that most hoaxes, like his own *Field Guide*, depended for their effect on the appropriation and subversion of the conventions of a particular genre. I was reminded of James Fredal's insight that the hoaxer "authors a fake that reveals a deeper fake and so by lying conveys a higher truth." Could the jackalope mount function as a critique of the violence, hypermasculinity, or anthropocentrism signified by the taxidermy trophy mount? Or were Andy and I overthinking this?

"Maybe it's a commentary on death as well," Andy speculated. "Taxidermy is weird that way. Kind of bridging that gap. You're not only preserving a jackrabbit, but now you've pushed it even farther into some kind of ethereal plane by preserving something that's not real." There followed a long pause in our conversation. "But you don't want to think too much about taxidermy," he said, chuckling. Before we said goodbye, I asked my final question.

"Andy, why do you think people love jackalopes?"

"What's great about the jackalope is that it's strange, but it's not scary," he answers. "It's weird, but it's kinda cute. And to make one is exciting because you're putting your hands on something you've lost touch with."

As a writer of books and a lover of field guides, I am smitten with Andy Robbins's jackalope hoax. But hoaxes—jackalope or otherwise—inhabit the wilderness of the Internet more often than the shelves of the bookstore. This is why many folklorists have turned their attention to how stories are originated, shaped, and transmitted within digital environments. Prominent among these scholars is Trevor Blank, whose fascinating work on Internet legends such as the Slender Man helps explain how nonhierarchical, decentralized online communities create contemporary folklore in a wide range of new media. Especially helpful is Blank's observation that "folk culture in the Digital Age is a product of hybridization. That is to say, new media technology has become so ubiquitous and integrated into users' communication practices that it is now a viable instrument and conduit of folkloric transmission; it works reciprocally with oral tradition." Blank's concept of hybridization extends to what he sees as the close relationship between folk artifacts, like a jackalope mount, and the dynamics of online vernacular storytelling.

Responding to my questions about his work on Internet folklore, Blank emphasized that the digital transmission of legends augments face-to-face forms of transmission rather than replacing them. He even remembered the jackalope being a subject of "xeroxlore" (also known as "photocopylore"), a sure sign that the technology of transmission is less important than the social function of folk stories. "There are a lot of uncertainties in life, disappointments, and things we don't fully understand," he told me. "Crafting stories, even mythical ones with seemingly outlandish components at times, helps ground us by connecting our tales with other people. Sprinkle in unverifiable components and the traditionality of the story and you have the makings of a long-lasting piece of folklore."

Blank also pointed out that the materiality of the jackalope mount helps the jackalope hoax to function. "Having the physical mount is the ultimate 'proof' to someone who isn't wise to the hoax. Jackalopes seem inherently funnier precisely because they've been 'documented.'" I asked him why he thinks jackalopes have been such an enduring part of American folk legend. Blank's first answer was "Because they're fun!" His follow-up reply focused on the issue of plausibility. "Another key part of crafting a legend," he said, is that "it can be supernatural, weird, or different, as long as it seems like there's some possible way that truth is in there somewhere. We like to pretend, make-believe, or even hold out hope that something truly otherworldly is lurking out in the dark."

While numberless jackalope stories and images may be found online, the horned rabbit has one particular digital habitat that I have frequented for many years: The Jackalope Conspiracy, a website created in 1997 by Richard Reibel. The Jackalope Conspiracy (whose delightful tagline is "Know the Truth. Live with the Fear.") not only asserts that the jackalope exists, but also that its existence is the subject of a massive government cover-up. The site offers such "evidence" as an obviously manipulated photograph in which US government scientists perform an autopsy on a horned hare and an entirely unpersuasive interview with a disgruntled employee from Area 51. It also documents the presence of jackalopes in ancient jewelry, religious idols, cave paintings, and hieroglyphics—all offered as indisputable proof that the jackalope has been living among us since antiquity. There is even a gallery of classic paintings in which the jackalope unaccountably appears: a jackalope joining Leonardo Da Vinci's *The Last Supper*; James McNeill Whistler's *Whistler's Mother* with a jackalope mount on the wall; Grant Wood's *American Gothic* with a jackalope peering out from the barn; Norman Rockwell's *Triple Self Portrait*, in which the painter leans to gaze into the mirror and sees not only himself but also a jackalope.

I was curious about what motivated Reibel to put so much energy into this bizarre website, but even more fascinated by the fact that so many strangers had been motivated to help him develop it. By far the most entertaining part of The Jackalope Conspiracy is its voluminous "True Stories" section, an impressive compendium of contributors' jackalope sightings and tales, sent in by folks from all over the world, not to mention from Nineveh, Romulus, and "Nexas [sic] of the Universe," wherever that might be.

The eclecticism of the jackalope testimonials is spectacular. A contributor concerned about jackalope roadkill offers this: "I am Sir Josef Ménageries, a servant to the royal Canadian King, from the Kingdom of Canada. I don't know why the US government does not put out warning signs on the roads for Jackalopes as they do for deer." Some respondents, like Tom, from Maryland, take a direct approach to proving the jackalope's existence: "I once had a jackalope hunting license. How could you have a license to hunt something that doesn't exist? The very idea is absurd."

Other testimonials have an outré streak. "Spanky," writing from "Earth," coolly explains that, while casually strolling in the woods, "looking for a more suitable host body," they spotted a jackalope holding a boom box and "clog dancing to the theme from Speed Racer." Someone from northern Maine tells a more disturbing tale. "I came home from work one day and found my son and his girlfriend having an orgy with a group of jackalopes. . . . And now he and his girlfriend (who is carrying the first human/jackalope) are planning to move to Canada to start their own colony. Don't let this happen to your child." Unsurprisingly, the thread of conspiracy runs strong. Holly, an anthropologist who relates a jackalope encounter in Nevada, concludes her narrative with this dawning revelation about the government's secret facility near Las Vegas: "I didn't make the Area 51 connection until I saw your site, but now everything is starting to make sense." Nathan C., from south Texas, takes conspiratorial theorizing to the next level: "I happen to know that the American Zoological Society knows all about the Jackalope problem, but has been forbidden by the government to tell the public for fear they might start a general panic. You see, the Jackalopes have been living and working among us for years, having the ability to change their appearance at will through the use of hypnotism."

Many of the tongue-in-cheek stories document the warrior rabbit's pure savagery. Sephra from Colorado tells the harrowing story of being "crippled for life" by a huge, snarling jackalope that attacked her in the forest, viciously biting her leg. "They had to amputate my leg," she laments, "but even worse, no one believes me." Worse still the fate of Chris G. from Bristol, Pennsylvania, whose sister was killed by a jackalope. Worst of all the suffering of Alley Gobblestone, from Indochina, whose mother was not only killed but also eaten by a jackalope.

The narratives submitted to The Jackalope Conspiracy are delightful, funny, wild, and occasionally disconcerting. Detailed first-person accounts prove beyond a reasonable doubt that jackalopes play golf, herd cattle, terrorize shopping malls, race ATVs, snort cocaine, sing opera, steal cars, and, perhaps inevitably, run for public office. Jackalopes are covert agents of the CIA. They are black belts in Jiu-Jitsu, fighter jet pilots, expert marksmen, and professional wrestlers. They discover buried treasure. Play shortstop. Create rainbows. My second favorite entry, submitted by Andrew J. from Savannah, Georgia, reads, in its entirety: "Are jackalopes real or not? I need to know because I bet a lot of money." (If only we knew which side of the wager Andrew took.) My favorite submission, because it is at once hilarious and disturbing, comes from Jesse C. in Minneapolis: "I'm giving a speech on the jackalope for my English class, and I appreciate the reliable information this page has given me! Thank ya much!"

Other stories offer helpful guidance about how to find, capture, and cook jackalopes (Wanda reports that "they are not bad eating, actually"). Also common are notes expressing appreciation for the website itself, or for the person behind it. Vincent writes that "as a physician, I have treated a handful of jackalope maulings in Texas. . . . Sadly, there is little public education about the dangers posed to the public by Texas Whitetail Jackalopes. Maybe your website can take the lead in distributing this information and, with luck, reduce the number of serious injuries." Other grateful notes speak directly to the writer's own jackalope conversion experience. This note from Mikster, in Brisbane, Australia, is representative: "I entered this site as a skeptic, but the evidence I have seen here is incontrovertible. I am now a believer."

Enter Richard Reibel, the guy behind this madness. The highest compliment I can pay Rich is to say that The Jackalope Conspiracy website is to the jackalope what *Weekly World News* is to Bat Boy. When he answered my phone call I immediately recognized in Rich's voice the upper North nasal twang so common in Michigan, where "begs" are for sacking groceries. We didn't exchange more than a few words before he began laughing—a hearty chuckle that continued throughout our hour-long conversation.

"Back in the midnineties I was trying to learn HTML, so I just made up some websites," he explained, barely able to get his words out for the laughter.

> The best was when I got on Howard Stern's radio show! I did this really crummy cam called the Amazing Grass Cam. It just looked at a yard. Nothing happened. It got mowed once a week or whatever, you know. Howard was really interested in that. It was so worth it because at the end of the interview, Robin, the girl he does the show with, said she just didn't get it. She asked, "Why do you do it?" Howard goes, "It's like making a statement. There are so many bad things on the Web, but this is the worst." Yes! I was over the moon. Howard Stern said my website was the worst!

I asked Rich how he came to work on jackalopes specifically.

> I really wanted one of those conspiracy websites. Did you ever see that old site *Bert is Evil*? Bert and Ernie, from *Sesame Street*, you know. It was this whole conspiracy about how evil Bert is. I mean, they had mugshots and everything, and all the photos were black-and-white, so it felt evil too. It was great! I'm sure they got a cease and desist from Children's Television Workshop. Well, I wanted to do something like that but I didn't want to get sued. Then my brother gave me a jackalope for Christmas and I thought, *maybe these things are dangerous and evil!* I didn't see a lot of jackalope stuff out there, so I thought maybe this is my conspiracy page. So I just made up a bunch of crap.

Still laughing, Rich explained that he didn't like how cute and sweet the jackalope was usually made out to be, so he dedicated himself to promoting the idea that the horned rabbit was actually ferocious. "Until I gave jackalopes a reason to be evil," he boasted, "nobody had anything bad to say about them." In the early days he received lots of email correspondence from folks who wanted to know if the jackalope was real. "Of course it is," he would respond. "So keep your doors locked!"

Rich went on to say how much fun he's had with The Jackalope Conspiracy over the years. His favorite part, he said, was receiving so many wild submissions from other jackalope enthusiasts. He clearly enjoyed the collaborative nature of the enterprise, describing it as "sort of crowd funded," though the site received no funding and the contributions were creative rather than monetary. These days, Rich says, his site is included as part of school curricula, used to help teach students how to identify online hoaxes. For this reason he takes special pride in making sure his site contains no profanity, and for years he has been happy to correspond with kids who are developing the critical thinking skills necessary to spot online deception.

Rich is the sort of person I could talk to all day—the kind of guy who always gets you laughing and never hesitates to laugh along with you. Before

I let him go, I asked my final question of the man whose bizarre website has provided so much entertainment.

"Rich, why do you think people love jackalopes?"

"Who doesn't like a rabbit?" he replied, without hesitation. "Rabbits are just kinda cute, and the rabbit with the antlers is kinda weird. You don't want to cuddle a chupacabra. They eat goats, man."

Chapter 4
Mounting Enthusiasm

But when I make a good [taxidermy] mount I feel like I beat God in a small way. As though the Almighty said, Let thus and such critter be dead, *and I said, "Fuck You, he can still play the banjo."*

—Christopher Buehlman,
Those Across the River

Anomalous as the jackalope might seem, it is part of a venerable tradition of corpse preservation. The artful preparation and restoration of dead animals has a long history, even as the idea must rank among the most disquieting in an extensive catalog of macabre human inventions. Because we are fascinated with mortality even as we fear it, perhaps it is natural that embalming, mummification, taxidermy, and other artificial means of preservation would be perfected as a way to sustain the illusion of life beyond mortal limits. Does surrounding ourselves with dead animal bodies as a form of commemoration or art show respect for nonhuman life, or does it instead reveal a failure to admit the necessary ephemerality of those bodies—and, by extension, our own animal bodies? Do we try to arrest these decomposing bodies in order to somehow forestall our own encounter with the unknown?

The ancient Egyptians were the world's first experts in animal corpse preservation, with examples of their work dating to at least 2200 B.C.E. Their sophisticated methods of embalming and mummification employed salt, beeswax, tree resins, oils, spices, and other microbe- and water-repellant substances. What animals have been found entombed in the afterwordly strongholds of the ancient Egyptians? Easier to ask instead what animal they did not preserve, as archaeologists have discovered in ancient temple burials the mummified bodies of ducks, falcons, vultures, frogs, lizards, fish, monkeys, baboons, cows, bulls, gazelles, lions, crocodiles, and even hippopotami. The Egyptians also preserved their beloved domesticated dogs and cats, which accompanied their owners on the mysterious journey to the afterlife. In fact, the ancient ones were so solicitous of their felines, they mummified mice and rats to travel with the kitties to the world beyond.

Mummification and embalming do not offer a direct lineage to the jackalope, since the antlered bunny is the creation of taxidermy—a species of carcass preservation different from that practiced by the Egyptians, who did not remove the skins of the animals they attempted to immortalize. The word *taxidermy* is rooted in the Greek *taxis*, which suggests order or organization and is the foundation of the word *taxonomy*, and *derma*, which you've already guessed means "skin." A common form of prototaxidermy practiced for millennia by cultures around the globe is the removal and tanning of animal skins for use as clothing, shelter, or for other purposes, both practical and ritual. True taxidermy involves not only skinning an animal but reconstituting it by mounting its preserved skin over a stuffing of straw, rags, or cotton, or stretching it over a mannequin. Also known as a "form," this may be fabricated from clay, plaster, wood, wire, resin, or, as is now common, synthetic urethane foam.

Taxidermy has been practiced for centuries. The Royal Museum of Vertebrates in Florence, Italy, displays a mounted rhinoceros that probably dates back to 1500. A stuffed crocodile suspended from the ceiling of a cathedral in Ponte Nossa, Italy, is documented from 1534, while the Museum at St. Gall in Switzerland exhibits a crocodile that was stuffed a century later in 1623. Modern taxidermy dates to the early eighteenth century, when books on the practice began to appear and the preservation of specimens for European natural history museums and private collections became professionalized and profitable. The heyday of taxidermy was the Victorian era, when both amateur and expert naturalists became obsessed with the collection and display of everything that flew, ran, crawled, slithered, or swam. The first World's Fair—the celebrated Great Exhibition of 1851, held in Hyde Park, London—brought together taxidermists from around the globe, who shared impressive displays of stuffed animals presented more dramatically than ever before. Characteristic of this dynamic new work was English naturalist John Hancock's *Struggle with the Quarry*, a sensational piece of taxidermy art in which a heron, as it grasps an eel, is savagely attacked by a gyrfalcon. During the nineteenth century, taxidermy flourished on the American side of the pond as well. Charles Willson Peale's natural history museum in Philadelphia included mounted animals along with displays of bones, skulls, and, most famously, the first full reconstruction of a mastodon skeleton. Henry Augustus Ward's Natural Science Establishment in Rochester, New York, featured exhibits so remarkable as to foreshadow the artistic turn taxidermy would take in the coming century. For the first time, taxidermists were working as reanimators, strategically employing death to bring animals back to life.

We tend to think of taxidermy in two cultural contexts that do not easily coexist. The first is the preservation of animals for scientific study and display; the second, the exhibition of animals as hunting trophies. Depending on your ethical stance toward nonhuman beings, you might find one or both of these uses of the animal body objectionable—or useful, beautiful or ghastly. But a third approach to animal taxidermy has emerged more recently. "Rogue taxidermy" (also known as "alternative taxidermy" or "crap taxidermy") is a weird contemporary offshoot of already strange pop-surrealist art, an underground visual arts movement that emerged from Los Angeles in the late 1960s. Rogue taxidermists are mixed media artists who use conventional taxidermy materials and techniques in highly unconventional ways. Most work within the bounds of an ethical code requiring that no animal be killed for the sake of art—even as these artists are content to work with dead pets, salvaged roadkill, hunters' scraps, or the refuse of animal rendering processes. While the inaugural exhibition of rogue taxidermy occurred in Minneapolis in 2004, the movement has spread from pop art to urban maker culture more broadly. Sarina Brewer, one of the movement's founders, describes her work as an homage to the animal: "It's really to commemorate the animal and tell that animal's story or my story through using that animal's body." The title of an article in the *Guardian* expressed my own question perfectly: "Rogue Taxidermy: A Misunderstood Ethical Art Form or the Next Hipster Fad?" Or, I wondered, might it be both?

Rogue taxidermy must be seen to be appreciated—or, if not appreciated, at least wondered at. An iconic example of the form is a curio sculpture consisting of the torso of a kid goat sporting quail's wings and the tail of a carp. Another piece includes the body of a fawn adorned with the head of a rooster and the tail of a pheasant; the fowl-deer hybrid is being ridden by a giant rhinoceros beetle, which grasps in its pincers chain reins leading to a bit in the rooster's splayed beak. In one of Julia deVille's best-known pieces, a stuffed black kitten wearing an elaborate funereal headdress pulls behind it a tiny hearse. In her work, Kate Clark often employs sculptural elements that allow her to produce stuffed ungulates with disconcertingly human faces.

Other examples of rogue taxidermy offer oblique commentary on the cultural iconography of nonhuman animals, and perhaps on their exploitation. An example is Brewer's own piece called "Mother's Little Helper Monkey," which consists of a wild-eyed, fanged stuffed monkey accoutered with wings (an allusion to the flying monkeys in *The Wizard of Oz*?) wearing an organ grinder's cap (a visual reference to the subjugation of animals for profit?) and, unaccountably, grasping an empty cocktail glass (an insinuation that I should mix a drink while considering what else it might mean?). A clearer example of social commentary is offered by Australian artist Rod McRae's installation "Are You My Mother?" in which a stuffed baby zebra gazes up longingly at a blank, white wall from which an adult zebra trophy head mount looks down at it helplessly.

Taxidermy has served the interests of everyone from Renaissance naturalists to trophy hunters to experimental artists. If I really want to understand the jackalope mount as artifact—as a product of the art and science of taxidermy—I'm going to have to meet the folks whose livelihoods depend on fabrication of the horned rabbit. My plan is to hop the cheap flight from Reno to Denver and head north from there to visit with several of the most notable jackalope taxidermists.

Leaving Denver International in a crappy, off-brand rental car, I drive north on I-25 along the scenic Front Range of the Rockies, yet another western paradise straining under the pressure of unchecked development. I pass by beerful Fort Collins and up into windy Wyoming. Not even stopping for coffee, I blow past Cheyenne, where my dad was stationed at F. E. Warren Air Force base back in 1955–56. I roll through Chugwater and Glendo, and then, just before reaching Douglas, the celebrated Home of the Jackalope, turn east on State Route 18 to Lusk (elevation 5,020, population 1,543). From there I enjoy a beautiful drive north through the remote prairie, up US Route 85 through Deadwood, where in 1876 Wild Bill Hickok took a bullet in the back while playing poker in a saloon and thus immortalized the "dead man's hand" of aces over eights.

The flat open land is dotted with old grain elevators and draped with a thin veil of snow. Occasionally I notice pronghorn browsing the grasslands. Square bales of hay are stacked neatly into golden cubes the size of barns while the fields of sunflowers are so vast it would take all of major league baseball from spring training to the fall classic to spit their seeds. One isolated ranch house is proudly flying the American flag and, next to it, a much larger, bright yellow "Don't Tread on Me" flag, its familiar coiled rattlesnake rippling in the wind. Train cars heaped with grain rattle by on tracks paralleling the road as I slow to pass a huddle of state troopers, each clutching a steaming styrofoam cup, clustered near a Budget rental truck that has flipped over and slid into the ditch. Towns are few along this route, though I notice at one crossroads the Sonrise Church and, at a second rural intersection, the Ammo Shack. A large billboard reads "Good Luck Hunters." A much smaller sign, offering a hotline number, addresses the epidemic of firearms suicides in the rural West.

After stopping for the night in Spearfish, South Dakota, to share some beer and time with an old friend, this morning I'm eastbound on I-90, passing through Sturgis—best known for the massive motorcycle rally held here each summer—on the short drive to Rapid City. Although Rapid City is the second-largest community in South Dakota, its population of 75,000 wouldn't fill some football stadiums. This morning I'll be meeting with Frank English, whom I've heard may be the most prolific jackalope maker ever to stick antlers on a bunny.

As I exit the highway and follow directions the taxidermist has given me, I'm surprised to find myself navigating a neatly kept suburban neighborhood, and I wonder if I've made a wrong turn. South Dakota is a decidedly rural state and I'm here to interview a guy who has reportedly fabricated thousands of jackalope mounts. I'd assumed I was on my way to a farm or ranch, not a prim row of split-level suburban homes resting neatly on uniform swards of green turf. But as I pull up in front of just such a home, the name on the mailbox confirms that I'm in the right spot. I hypothesize that while English lives here, his jackalope-making operation must exist offsite in a picturesque old barn somewhere out on the windswept prairie.

Frank English, who appears to be in his midseventies, answers my knock, invites me in, and leads me from the foyer of his split-level home up into the living room, where he introduces me to his wife, Dianne. He shows me to a large chair and offers me a bottle of water before seating himself on the couch across from me. On the phone Frank had been reserved, but now he launches comfortably into stories. He's from the small town of Havre, up in Montana, and he spent forty years in the military before being assigned to his final post at Ellsworth Air Force Base here in Rapid City. A few years before he wrapped up his career with the Air Force in 1981 he took up taxidermy as a hobby to have something to fiddle around with in retirement. As he became more adept at making jackalope novelty mounts, he sold a few to local businesses, to the airport, and eventually some to Wall Drug, the famous purveyor of kitsch just an hour east of here near Badlands National Park.

"Then, thirty-five years ago, I got the call." Frank grins.

"The call?" I repeat.

"Yup, and I'm gonna tell you all about it, but you're not gonna put any names I tell you in your book. Agreed?"

I give Frank my word, close my field journal, and sit back as he tells me the incredible story of his recruitment by an outdoor equipment retailer who tapped him to make jackalope shoulder mounts for sale through their catalog. His first order, he says, was "a small one, just three hundred."

"I'm sure I still have that first invoice. I keep all our receipts," Dianne adds, heading downstairs. I hear the sound of a metal file cabinet drawer clicking closed and she returns momentarily with a handwritten invoice dated from December 1984 in which the agreement to produce three hundred antlered rabbits is spelled out in terms I agree not to disclose.

"That was the start of it all," Frank beams. "Some people didn't understand that the jackalope was here to stay. They figured it was a fad, but I knew better. You know what else was supposed to be a fad? The skateboard. Kids used to make them out of plywood and roller skate wheels. That first three hundred was just the start of it. I've been making them—me and Dianne—ever since. Thirty-five years."

Frank stands up, smiling. "You ready to see where the magic happens?" I nod, expecting that we're about to jump into a pickup and head out onto the prairie. Instead, Frank simply walks downstairs, turns right at the foyer landing, and descends to the lower level of the house.

As I follow him down a few steps into an unfinished basement, it is obvious that something very strange is going on here. The floor is covered with coarse indoor-outdoor carpeting while the walls and ceiling reveal the exposed studs, which have batts of insulation stuffed between them. Against the foundation wall are a desk, office chair, and file cabinet. Dangling above are two large mounted geese, posed dynamically to suggest a landing. On the wall are several large display boards, each of which features nine jackalopes sporting improbably large antlers.

"These fellas are my famous World Record Jackalope mounts. I invented them. Fourteen-point bucks! But I've invented a lot of special ones. I was the first guy to make a mount with the jackalope turned sideways, like this." Frank turns his face away from me to demonstrate the profile pose. "And I came up with a mount that was a little pair of buck and doe jackalopes together. Girls really loved that one! And guys ate up the one I used to make of two 'lopes fighting each other. Those were all specialty mounts, you know. They take a long time to make. Once I even made a jackalope that I dressed to look like a hippie, and he had really red eyes and he was smoking a joint. That one was cute as hell! I tried to go back to it, but they were going to kick me off eBay." Frank's enthusiasm is so palpable that I find myself trying to imagine what a stoned jackalope might look like.

"You know why I have my shop down here?" he asks. "So I can do my work without my proprietary techniques getting stolen. I'd *never* show anybody how I make a jackalope. Well, I taught my son. He's a taxidermist, too, but he does the big stuff. You know, bison and stuff like that. Anyways, I'm always trying to improve the product. Lots of innovation."

Dianne chimes in quietly. "I'm his quality control. Frank works with the forms and he's the skin stretcher but I glue the eyes and pin the ears. It has to be done just right." Dianne touches the tip of her pointer finger to her thumb as deftly as a surgeon to suggest the precision of jackalope ear pinning.

I ask Frank how his jackalope making has changed over the past thirty-five years, and he lays out the history of his jackalope operation:

> Well, in the beginning I shot all my own jackrabbits. But I just couldn't keep up, so I hired some boys to do my shooting. Just put an ad in the paper, you know. But I only hire guys who

> take good care of my rabbits. I'm very particular. The jack's gotta be taken in January, February, or March. Outside about a two-and-a-half-month window the fur isn't top notch. I'm particular about the shooting too. When I hire the boys I specify the caliber of the rifle, and I won't buy a rabbit that's seen the business end of a shotgun. Some things you just can't patch. Took a while to get it all figured out, but it runs real smooth now. I also hired a skinner, so I just get the skins now, no field dressing at all. And you gotta stay ahead of the ups and downs in the rabbit population. Some years there's a lot more out there than others. That's why I stockpile skins. Production would stop if I ever ran out. We came close a few times.

Frank looks over at Dianne, who nods in agreement.

"Come look," Frank says, leading me to the far end of the basement, where he shows me a large white top-loading freezer that is humming away. "Go ahead, open it," he instructs, smiling. I grasp the chrome handle, lift the heavy lid, and find myself peering down at hundreds of plastic grocery bags, each one containing the frozen, bloody pelt of a rabbit.

"Six hundred seventy-five in there now, but I've got twice that sometimes," Frank says, proudly.

Jackalopes are cute. They're funny. They're a delight. Few people can see one without cracking a smile. But now that I was witnessing how the rabbit sausage was actually made, I worried that the comical antlered bunny might have lost its innocence forever. Of course I knew that the fabrication of a jackalope mount requires the skin of a rabbit, but it was another thing to confront hundreds of bloody pelts in a basement freezer. I closed the lid slowly and attempted weakly to feign the role of dispassionate journalist.

"What about the antlers?" I asked.

"Oh, I used to get horns from butchers all over. There's nothing else they can do with the horns, so they liked it. They made money off me. Hell, I had horns shipping in from as far away as Wisconsin, even Texas. But then you can't believe what happened. Customers started complaining that the horns didn't match perfectly. Didn't like that they were different! Anyway, about eight or nine years ago I got it all worked out and now I have them made in China. They do it the same way they make dentures. Very exact. Now they crank 'em out for me, all the same. Want to see?"

Frank leads me to a small door that opens to a storage space beneath the main stairway of the house. He flips on a light and I bend over to have a look. Beneath the stairs are eight or ten large plastic storage bins brimming with small antlers—all identical, each indistinguishable from the rest. I felt a mild sense of disappointment in seeing these precisely fabricated "antlers," which seemed to defy the unique spirit of a jackalope. Just moments before I had gazed into a freezer full of bloody rabbit skins and been troubled by the slaughter required to produce the good fun that is a jackalope mount. Now I was looking at hundreds of fake racks and thinking that the use of imported artificial "antlers" had crossed an invisible line between craft-based American folk art and homogenized global mass production. Both the real rabbit skins and the synthetic deer antlers troubled me, but my own hypocrisy was obvious. I simply couldn't have it both ways.

"So that's the skin and the horns. But the body of the jackalope is a foam form that's made from a fiberglass mold. I make my own molds. Every one. Some of them are thirty years old now, and there are secrets to doing it right. That's all proprietary. But once I get the set-up done the way I like it I can use help, so I have a woman who works from home and makes a lot of my forms." Frank guides me to a small shop off the main room. Here he has a long workbench with good lighting and a large magnifying glass on a flexible arm. Various species of mounted fish adorn the walls. Frank notices my eyes panning. "Yeah, I started on fish. This was my first one," he says, leaning forward to closely examine his own work on a black crappie. "Not too bad, really. But I'm a lot better now."

Frank reaches into an adjoining utility closet and pulls from a high shelf a blue, plastic bin the size of a shoebox. He shakes it like a maraca and smiles at the sound. "You know what's in there?" I pause, not quite recovered from my experience with the freezer. "Go ahead, open it up."

I should have guessed: hundreds of eyes. "Changed my whole operation when I figured a way to make my own eyes. And I'm gonna show you how I do it, but you're not gonna put any of this in your book."

I nod my agreement, clicking my pen shut and sliding it behind my right ear. Frank goes on to demonstrate the technique he employs to make the eyes, which gleam like ebony. "I only use black eyes. I've tried other colors, but customers felt like the 'lope's eyes would follow them when they crossed a room and that creeped them out. Black only now. Always improving the product." I'm genuinely impressed by Frank's resourcefulness and ingenuity, and by the obvious pride he takes in his operation. Facebook was launched in a Harvard dorm room, Hewlett-Packard founded in a Palo Alto garage. Here, in the basement of their suburban South Dakota home, Frank and Dianne had developed a highly efficient system of jackalope fabrication.

"Lots of innovation going on down here," Frank says, proudly. "You have to work hard if you want to make something of yourself. I even came up with a process to keep the fur out of your nose while you're working with it. But that's proprietary," he adds. "So that's the skin, form, horns, and eyes. All that's left is the wooden plaque I mount it on. I've got a guy in Sturgis who makes my plaques. He likes the job since he can make them whenever he wants and it gives him more time to hunt."

"So, I've showed you how I make a jackalope," Frank says, smiling. "But I haven't showed you how I make a jackalope. I'll never teach anybody."

"You've been making these guys a long time," I observe, scanning Frank's eighteen World Record Jackalopes staring back at me from their neat rows on the basement wall. "You must have some stories by now."

"Oh, sure," Frank nods. "There was the time I got a call from a lady in England who took one of my jackalopes with her to Australia to give to her friend. They confiscated it at customs! Thought it was an endangered species. Took it away and put it with the elephant tusks and leopard skins. She called and asked me to write a letter explaining all about the jackalope. I put it on letterhead and everything. It took a while, but I got it back for her," he chuckles.

"And here's something that used to happen all the time. I mean *all* the time. When I started out I would put a little sticker on the back of the plaque with my name and phone number. Figured it could lead to more business. Of course a lot of my jackalopes ended up in bars. Well, guys would get to drinking and somebody'd be trying to pull one over on somebody. Then there'd be an argument about whether jackalopes are real. And somebody'd always get the bright idea of calling the guy who made it to settle the fight. Well, hell, I mean these mounts are all over the place, so I was getting calls from everywhere. California, Florida, Ohio. Usually middle of the night, you know, drunks calling me up."

Now it's my turn to laugh. "What would you tell these guys?"

"Yes, I'm the taxidermist who mounted it. Yes, jackalopes are real. Now go drink some black coffee and don't ever call me again."

Before I wrap up our conversation, Frank agrees to let me take a few pictures, but he's specific about what's off limits. He allows me to photograph his custom jackalope form, but only from an angle that won't reveal one of his many craftsman's secrets. Before I wrap up my photo shoot Frank says, "You have to get one more. Check this out." He takes one of the fourteen-point World Record Jackalopes down from the wall and walks it to the far end of the room. Here he shows me his custom-made jackalope photo-op station, complete with directional lighting and, mounted on the wall, a large display board neatly upholstered in black velvet. In the middle of the fancy board is a special hook on which Frank hangs his jackalope. He snaps on the light and directs it perfectly to eliminate shadows. The trophy buck is ready for his glam shot, and I click away.

Putting my camera back in its case, I pick up my journal and ball cap, and thank the Englishes for their time and hospitality.

"You owe me a book," Frank says.

"Deal. But it might take me a while," I answer.

"You just have to find ways to make it more efficient," he advises. "Keep improving your product. You have to work hard if you want to make something of yourself." It's an odd way to talk about writing, but I suspect Frank is exactly right.

"Frank, how many jackalopes have you made?" I ask.

"Total, you mean? Oh, hell, I don't know." He looks at Dianne, who I can tell is already doing the math.

"Two hundred thousand," she says.

"Thirty-five years," Frank says. "Yeah, at least two hundred thousand. I'd say more."

Dianne nods in agreement. "It could be more. But two hundred thousand anyway."

"And every last one of them was made right here," Frank says, as he looks around his basement. I shake his hand and thank him again.

As I step out onto the front stoop, I turn to ask my final question.

"Why do you think people love jackalopes?"

Frank answers first. "A lot of guys got fooled with the jackalope by older guys when they were young, and now they're older themselves and they want to fool the younger guys. It's a generational thing."

"It's just a nice comical gift for somebody special," Dianne adds. "Something nobody has."

I climb into my rental car and offer a final wave as Frank English swings his front door closed. I'm stunned by the sheer number of mounts he has fabricated over the past thirty-five years and the variety of techniques he has developed along the way. I now know without question that many of the jackalopes I've seen around the country began their migration right here. I've just met the Benjamin Franklin of jackalopes: a man whose dedication to innovation and efficiency has transformed the basement of a suburban house into the least likely and most productive jackalope factory in the world.

Today's journey will take me 250 miles from one jackalope maker to another. Leaving Rapid City, I head west and double back through Sturgis and Spearfish and on to Gillette, Wyoming, where I turn south on I-59 toward the Thunder Basin National Grassland. This is open country, a picturesque prairie landscape stretching out beneath a cloudless sky. It is also the epicenter of the earthquake that is the new American energy boom, and this basin is exploding. Almost every vehicle on the ribbon of road is heavy equipment making its way to remote drilling operations and fracking pads. In every direction the long view is spiked with pump rigs to the horizon line. If this massive operation is built out according to plan, there may be as many as five thousand new wells drilled over 1.5 million acres of this beautiful prairie basin. Six-thousand-year-old pronghorn migration routes will be impeded, vast heritage grasslands torn up, and irreplaceable habitat devastated, all to prolong the life of a carbon-intensive economy that is already ravaging the planet.

I press the scan button on the rental car's radio to distract myself from this wholesale destruction, but it is all static and preaching, preaching and static. Opting for silence, I suddenly feel homesick for the unfracked barrenness of my home desert. I pass through Wright, Wyoming, yet another town whose population is half its elevation. One more hour brings me back to Douglas again, where I catch a glimpse of the thirteen-foot-long jackalope silhouette adorning the ridge outside of town as I swing west onto I-25. Another hour and I'm passing through Casper, population 58,000, Wyoming's second-largest city behind Cheyenne. My destination is Mills, a community of four thousand resting quietly on the western edge of Casper, not far south of the township of Bar Nunn.

Michael Herrick wasn't at all talkative on the phone, so I don't have a good read on what kind of person I'm about to meet. I do know that, as a jackalope craftsman and the son of Douglas Herrick, the jackalope's inventor, he is the living link to the jackalope's Depression-era origin story.

Rolling into Mills on West Yellowstone Highway, I can't miss a large sign reading "AS REGULATIONS GROW, FREEDOMS DIE." I pass Poison Spider Road and the Powder River Armory, with its advertisements for guns and ammo, then a Family Dollar strip mall and the Buckin' Brew Restaurant. The whole town has an edge-of-town feel. At last I pull into the dirt parking lot of Herrick's shop, Antler Taxidermy and Arts, a metal building situated across a potholed alley from a small, open-sided game processing facility where hunters bring their kill to be butchered and packaged.

The door to Herrick's small business is on the side of the building facing the game butcher's shop, where I notice two deer carcasses dangling. As I step inside Antler Taxidermy and Arts, I am immediately struck by the remarkable size and dynamism of the animals on display. Every wall is packed with big game shoulder mounts, including deer, pronghorn, elk, even moose. Against one wall stands a black bear, upright on its hind legs, forepaws raised and claws outstretched. In a corner of the room is a full-body mount of a bighorn sheep. While there are a few alien interlopers—I recognize a three-foot-long alligator gar from my days paddling and fishing in the Florida Everglades—this is primarily a museum-quality exhibition of mammoth specimens of Great Plains wildlife. The centerpiece of the display is the full-body mount of a mountain lion that stands fiercely in the center of the room, making it impossible to enter the shop without experiencing the uneasy feeling that the big cat is preparing to pounce. Its small ears protrude from its head and the humps of muscle domed above the shoulder blades bespeak inhuman power. The panther's head is lowered slightly, as it begins a crouch that winds tight every sinew in its body. The cat's eyes are piercing and its long tail appears midtwitch, the tic of that tail the last thing you might notice before being lost in the blinding flurry of the attack.

I am no hunter, and I'm uncomfortable with the ethics of trophy mounting. But as I gaze around this shop, I'm impressed by the precision and beauty of these mounts and captivated by how dramatically they bring the animals to life in an act of imagination that is as thrilling as it is disturbing.

Backing away from the mountain lion, I move slowly toward an interior door that stands slightly ajar, appearing to lead to a back room.

"Mr. Herrick?" I call out over what sounds like a fan blowing from behind the door.

"Come on back."

Swinging the door open, I enter a well-lit workroom with the gleam of a laboratory, across which a man is bent over a working table. He is wearing boots and jeans, an old work shirt, a green-and-white tractor cap, and smallish glasses which, as he looks up, ride low on the bridge of his nose. He appears to be in his early sixties. In his hand is a blow dryer.

"Drying glue," he explains succinctly, switching off the power and pointing the barrel of the hairdryer toward the ceiling as if he's brandishing a Colt revolver. Although there is a six-foot elk rack standing directly between us on another worktable, a glimpse confirms that the man has been focused on a jackalope-in-progress. Behind him an exterior door stands open and through it I see the decapitated head of an elk lying in a dark pool of blood in the gravel.

Setting the hair dryer down, he walks slowly over to me, extending his hand.

"Mike," he says.

"Same here," I reply, shaking his hand.

As Mike Herrick walks past me and into the display area of his shop, I can only assume I'm supposed to follow. He sits down at a small desk chair in a corner of the room across from the bighorn sheep and looks up at me in silence. There is no phone or computer on the desk and, since there is no other chair in the room, I stand waiting awkwardly. After a silence during which each of us is as uncomfortable as the other, he finally speaks a single word.

"Jackalope." This doesn't seem to be a question, exactly, but I take it as my cue to state my business. I glance at my notes and, in a fumbling attempt to start a conversation, ask about his earliest memories of jackalopes.

"No memories before jackalope," he says, with a slight drawl. "Was always jackalope. My dad taught me. His way was a whittled-down two-by-four in the head, covered with papier-mâché. Good mold. Nobody makes 'em like that anymore. Different forms now. Different preservatives and glues."

"How long have you been making jackalopes?"

"Was born right here by the river. Started mounting jackalope with my dad at thirteen."

Thinking of the prolific jackalope maker up in Rapid City, I ask how many jackalopes his father made before his death in 2003.

"Dad invented jackalope, but never did make many. Most maybe fifty a year. Less, usually. I make about that now, a few more. My cousin Jim over in Douglas makes a lot. Can remember when dad sold jackalope for twenty dollars. Mine go for one hundred twenty dollars. That's just what it takes . . ." his voice tapers off, modestly, though I imagine he'd like to say, "what it takes . . . *to do it right*."

The only way to get information out of Mike is to ask direct questions after each of his brief answers is concluded. The result is a staccato rhythm that makes the conversation feel like a ping-pong match. I am still standing.

"So, your dad didn't make many jackalope mounts. Did he specialize in big game?" I ask, glancing over my shoulder at the mountain lion.

"Nope. Wasn't a taxidermist but on the side. Was a pipe fitter, welder at the Amoco refinery." I'm starting to feel a sense of connection with this other Mike, reserved as he is. My grandpa was a pipe fitter, too, in the Navy, and one of my earliest memories is of the anchor tattoo on his bulging forearm. "Dad had polio on his left side. Had to use his right arm to lift his left."

"I never read that anywhere," I say with mild surprise.

"Before pipes he was a tail gunner in the war." I recalled learning that Douglas Herrick had served on a B-17 in the South Atlantic Division, Air Transport Command, Brazil. There follows another long pause. "He gave something to the world when he invented jackalope. People make fun of it, but that's ok. It's part of the whole Herrick family heritage."

I'm starting to discern what was not clear to me at first. Mike is not crusty but instead sensitive, not inhospitable but introverted. Talking is a chore for him, and I sense that speaking of his father, who has been gone fifteen years, has brought up feelings of pride and sadness in him.

I change the subject to give Mike a reprieve. "Looks like you've got a lot of big stuff waiting for pickup," I say, sweeping my arm to indicate the twenty or so large mounts on display.

"Nope. I killed all that," he says, flatly.

I am taken aback by the idea that Mike hunted all these animals himself, right down to the mountain lion.

"Have jackalope mounts been good business for you over the years?" I ask.

"If dad had patented jackalope I wouldn't be here right now. But that's okay. He gave something to the world." I had always treasured the fact that the jackalope is part of the folk tradition, enjoyed by all and owned by none. But now, as I stood in the studio of a man I was coming to understand as a struggling artist, I wondered what even one percent of every jackalope mount, T-shirt, shot glass, and postcard sold in the past seventy years would amount to. It was a staggering contemplation.

"I've always liked making jackalope. I sell some, but mostly it's bigger game—when there's work. And I'm a sculptor." Mike draws my attention to an alcove in the back of the room that is adorned with delicate cast bronze sculptures of wildlife including bighorn, bison, and moose. The detail of the bronzes is exquisite, and the precise attention to the anatomy of each species depicted is flawless. "Nobody buys sculptures, but I'd cast every day if I could afford to," Mike says, a tone of longing dampening his voice. I wonder, momentarily, at the small fortune this spectacular work might fetch in a boutique gallery three hundred miles west of here in Jackson Hole.

"When you make jackalope mounts, do you hire anybody to do your hunting for you?" I ask.

"Nope. Do all my own shooting," he replies, shaking his head.

"How about skinning? Or making the forms? Or the plaques?"

"Nope," he replies, offering no elaboration.

"Do jackalopes have individual styles associated with their makers?" I ask. "I mean, if you're out somewhere and you see a jackalope, can you tell if it's one you made?" Now he stares blankly, as if I am decidedly the most ignorant person ever to have set foot in his shop.

"Yes."

"What if you saw one of your father's mounts?" Mike sits up a little straighter in his chair.

"Yes."

"I'd like to buy one of your jackalopes," I say suddenly, surprising myself. While I have a deep passion for scaling the mountain of jackalope narratives, kitsch, and art that Douglas Herrick's invention has inspired, I've never felt the need to hang a dead rabbit head—antlered or otherwise—on my wall. But something about the laconic Mike Herrick has changed my view. I had sensed a son's love for his father, seen an artist trying to reckon the value of his work. I had watched an American craftsman's hand making something unique without outsourcing, upscaling, or compromising. I had witnessed both the poverty and richness of this man's quiet determination to keep his father's legacy alive, one beautifully fashioned horned rabbit at a time.

"I'd like to buy that one you were working on when I came in, and I'd like to pay a buck twenty for it" I continue. "Could you send it over to Nevada whenever it's finished?"

"Sure thing," Mike answers. I take out my wallet and offer him my credit card. He looks at me awkwardly, and I realize that he has no way to process it.

"I'll send jackalope over soon as I get him finished," he says. "You just mail me a check."

I thank him, and then begin the question I've come so far to ask the son of this whimsical creature's inventor.

"One last question, Mike. Why do you think people love . . ."

At just that moment, four guys in full camo stream into the shop, talking loudly. Our quiet conversation is overwhelmed by the energy and excitement of the hunters.

A large man who looks about Mike's age slaps a much younger man on the shoulder. "This boy got himself a big buck today! We got it out in the truck and figured you could fix him up with a memory!"

"I can do that," Mike replies. "Just finishing up with this fella and then I'll get right with you."

Mike leads me back into his studio, where I notice the hair dryer lying next to the unfinished jackalope. He offers me his hand. As we shake, he nods his head toward the hunters waiting in the front of the shop and says, "That's a lot of groceries. You understand."

"Sure thing, Mike. Thanks for your time." Leaving the building I overhear the men asking about the mountain lion, though I can't make out Mike's quiet reply.

I have connected with the son and protégé of the jackalope's inventor, and I will now have in my home a piece of genuine American folk art that will bring me as close to the jackalope's origins as I'm likely to get. As I dodge potholes while crossing the gravel parking lot, closed journal in hand, I see the black-eyed buck splayed in the bed of the hunters' mud-splattered pickup.

Chapter 5
A Mighty River of Jackalopiana

So when you're searching for the truth
and you're at the end of your rope,
You might find you don't need no proof
to believe in the thing that gives you hope,
And for me, that's the jackalope.

—Grammy Award–winning American roots musicians
the Okee Dokee Brothers, "Jackalope"

The jackalope mount is the headwaters of an endless stream of gewgaws, knickknacks, bibelots, doodads, tchotchkes, and trinkets that has flowed unabated through popular culture since the 1930s. For example, a quick search for "jackalope" on Etsy, the online craft clearinghouse, turns up a veritable warehouse of cheap jackalope crap: tote bag, necklace, earrings, stuffed animal, pint glass, coffee mug, shot glass, throw pillow, sticker, pin, wall poster, Halloween decoration, Christmas ornament, birthday card, pet toy, cross stitch, jigsaw puzzle, pocket mirror, leggings, bandana, ball cap, hoodie, sweater, belt buckle, panties (complete with "Horny Rabbit" emblazoned on the crotch), bib, apron, weathervane, snow globe, refrigerator magnet, welcome mat, golf balls, playing cards, luggage tag, toilet lid cover, key chain, T-shirt, and oven mitt, just to name a few.

Scholars in various fields—including art history, philosophy, architecture, and cultural studies—have long debated what constitutes kitsch, and it may be the holy grail of aesthetics to explain why we are so drawn to it. According to the earliest use of the term, which emerged in Germany in the mid-nineteenth century, kitsch is failed art that is superficial, imitative, sentimental, or banal. It is bad taste commodified and then consumed by those too ignorant to know better. While the concept of kitsch did reliable service for critics desperate for a way to distinguish what they considered real art from mass-produced failed art, in a post–Andy Warhol world any firm distinction between art and kitsch has been severely troubled. The perennial appeal of kitsch is more easily demonstrated than explained. In their book *Kitsch! Cultural Politics and Taste*, Ruth Holliday and Tracey Potts observe that "kitsch seems to be in possession of some kind of irresistible vital force, despite protestations to the contrary." Or, as Milan Kundera put it in his 1984 novel *The Unbearable Lightness of Being*, "None among us is superman enough to escape kitsch completely. No matter how we scorn it, kitsch is an integral part of the human condition."

Jackalope junk is often mass-produced and tasteless, but is it aspirational enough to be considered kitsch? Even the wonderful Museum of Bad Art (MOBA) in Boston maintains that terrible art can be produced only by those who attempt to produce good art but fall (very) short. As one MOBA curator commented, "We collect things made in earnest, where people attempted to make art and something went wrong, either in the execution or in the original premise." But how earnest was the artistic attempt made by the producer of my flimsy jackalope bottle opener? Do my daughters love their jackalope keychains because the makers of those objects set out to produce great art?

Perhaps jackalope junk should instead be associated with camp, a sensibility distinguished by a love of bad art not in spite of its badness but because of it. As Susan Sontag observed in her influential 1964 essay,

"Notes on Camp," "Camp is a vision of the world in terms of style—but a particular kind of style. It is the love of the exaggerated, the 'off,' of things-being-what-they-are-not." While often associated with queer culture, commonly supplied examples of camp run the gamut, from camp icons Liberace, Cher, and Lady Gaga to films like *The Rocky Horror Picture Show* and John Waters's *Pink Flamingos*. Intentionally eccentric and gaudy, camp celebrates the deliberate excess and tawdriness associated with lowbrow cultural production of all kinds. For that reason, camp is often distinguished by a delightfully unstable alloy of honest pleasure and pure irony, and usually functions as an attack on the pretensions of highbrow culture. For example, I know a guy who earned a PhD from Harvard and who proudly displays a framed print of the awful (and thus immortal) painting *Dogs Playing Poker* over the mantle in his living room. Think of it this way: if you attend an ugly sweater–themed holiday party wearing your grandma's best effort, that's kitsch; if you instead arrive at the get-together in a sweater purposefully fabricated to be garish and tacky, that's camp.

Although the jackalope taxidermy mount is the most iconic representation of the horned rabbit, there are many more horned rabbit T-shirts, keychains, and shot glasses than there will ever be mounts, and so the jackalope's wide distribution in popular culture is attributable more to that ever-flowing stream of jackalope junk than to the work of any taxidermist, however prolific. The earliest and most recognizable of these tacky souvenirs are jackalope postcards, which have been circulating since the Great Depression, though extant examples from before the 1940s are about as rare as the strange rabbit itself. Almost from the moment that commercial photography, mass stock printing, and new postal delivery protocols combined to make the postcard a commercially viable mode of communication in the early twentieth century, "tall-tale postcards" were popular as a visual analogue to the narrative tall tale. Cards depicted flatbed trucks loaded with ears of corn the size of redwood trees or potatoes so immense they had to be hauled by teams of Clydesdales. Men were seen riding bullfrogs the size of actual bulls or wrestling man-size grasshoppers while monster pike or bass seized fisherman by the legs, dragging them from their boats. Among these hyperbolic visual representations of the immense scale of the American land, rabbit depictions were especially popular. In a card issued in 1935 by Stovall Studio in Dodge City, Kansas, a jackrabbit appearing to weigh several hundred pounds is being hoisted by three men using a block and tackle. Fitting nicely into this rich genre of tall-tale postcards—which are a clear precursor to photoshopped images—are cards depicting fantastic species, including the fur-bearing trout (or "beaver trout"), the Wisconsin hodag, and, king of them all, the jackalope.

Vintage tall-tale postcards are housed in archives around the country, but the largest private collection of jackalope postcards, consisting of a remarkable 276 distinct cards, belongs to Utahan Sean O'Brien, who has generously made available online images of almost one hundred of his jackalope cards in what he calls the Jackalope Postcard Museum. The sheer variety is astounding. Sean has collected cards from Wyoming, Montana, Arizona, Utah, Colorado, New Mexico, Texas, Oklahoma, South Dakota, California, Oregon, Kansas, Nebraska, Michigan, and as far afield as Canada and Germany. In one, a cowboy has saddled up and is herding cattle while riding a giant jackalope. In another, a Texas jackalope sports longhorns in place of deer antlers. There are jackalopes posing before natural wonders, including the Grand Canyon and the Grand Tetons. Other jackalope cards promote businesses: the Elk Horn Café in Sundance, Wyoming; the Buffalo Inn Restaurant in Goodland, Kansas; the Club Café in Van Horn, Texas. Sean possesses many cards still bearing handwritten messages from correspondents now long gone, including some postmarked with a single one-cent stamp, a postal rate that existed only until 1951. His collection includes jackalope cards by excellent professional photographers including Harold Sanborn, and many by the king of novelty postcards, Bob Petley, who launched Petley Studios in Arizona in 1943 and by the late seventies was selling forty million cards a year (not all jackalopes, of course). And while vintage cards hold the greatest historical interest, jackalope postcards are still being made and mailed today. My favorite among this more recent work is a series of delightful cards by Austin-based artist Chet Phillips—cards variously depicting a friendly jackalope tubing the river, paddle boarding, and jamming on guitar in a bluegrass duo.

When I called Sean O'Brien to learn more about his extraordinary jackalope postcard collection, I expected to talk with a person who, like myself, is fascinated by the horned rabbit. But in addition to being possessed by the jackalope, Sean is also a serious collector, and there is a reason why collecting has sometimes been compared to addiction. Sean told me that his collection began innocently with a pair of jackalope postcards his mother brought him from a trip back in 2004. "I never set out to have the world's largest collection of jackalope postcards," he told me, though by now he certainly does. For almost twenty years he has been on an obsessive mission to identify, locate, and acquire every jackalope card in existence—an impossible goal that is thus all-consuming. He has found cards in gift shops, thrift stores, and through antique dealers. He has bid for and purchased them from a variety of online sources and has even collected a few by going directly to postcard manufacturers. Sean even plans family vacations with an eye to visiting spots where he might turn up a new card. But in doing all this, he confronts the crucible faced by all collectors: the more cards he acquires, the more difficult it becomes to unearth new ones, and the more driven he becomes as a result. Sean told me that he maintains a bank of around thirty images of cards that he knows exist but has never been able to add to his collection. Thinking about those cards still out there, beyond his reach, drives him to distraction. Collecting, Sean observed, "can kind of take over your mind and make you miserable." That prompted me to ask if he thought the frustration might ever cause him to give up collecting. Not only did he have no intention of quitting, but his ambition is to develop his collection further and someday donate it to the Smithsonian.

Where, on the surprisingly fluid art-kitsch-camp continuum, should we place Sean O'Brien's 276 postcards and the other numberless items of jackalopiana produced since the 1930s? I would certainly argue for Mike Herrick's jackalope mounts as art, though anyone who prefers Rembrandt might reasonably disagree. I likewise assert that the labor-intensive, handcrafted, one-of-a-kind and yet incredibly ugly orange-and-green jackalope afghan I saw for sale on Etsy is kitsch, because I believe in my heart that the maker of that abomination had high hopes and did their level best. But what of the awful snow globe depicting two jackalopes holding each other's paws while exchanging wedding vows? I can only hope that stunning artifact is pure camp—a mass-produced object that revels in the glory of its own tackiness, replicability, and open defiance of sincere artistic ambition.

However we might taxonomize jackalope junk and our aesthetic responses to it, it is clear that Sean O'Brien and I are not alone in loving horned rabbit kitsch and camp. When my family asks why I have purchased yet another jackalope T-shirt, I can muster no adequate reply. For example, I recently bought online a putrid urine-yellow T-shirt depicting a very poorly drawn jackalope doing what looks like an awkward dance (or might he be having a stroke?) in front of the splayed pads of a large prickly pear cactus. I was overjoyed when it arrived. "Dad, that thing is awful!" my daughter, Caroline, exclaimed. "I know," I replied, with a delight at once deeply ironic and perfectly genuine. "Isn't it fantastic?"

Whether we consider jackalopiana to be art, kitsch, or camp, what is most impressive is how much of it there is and how long it has remained popular with a wide audience. And no single establishment has done more to disseminate all-things-jackalope than South Dakota's celebrated Wall Drug Store. As an aficionado of jackalopiana, I felt it essential that I make a pilgrimage to what is unquestionably the world's most famous jackalope emporium. Carving out a few days after attending a convention in Boulder, I retraced my earlier route up I-25 from Colorado through eastern Wyoming, skirting western Nebraska and again doubling back through Spearfish and Rapid City. Not long after crossing the South Dakota state line I began to spot the celebrated Wall Drug road signs which may be seen for hundreds of miles in all directions on every route approaching the little town of Wall (population 814). This ad campaign began back in 1936 and the signs remain a delightful throwback to a bygone era: each is made of wood, hand-painted by local artists, and features anachronistic messages like "Wall Drug: Free Ice Water." I can only imagine what a novelty that cold water would have been during the Great Depression, when Wall Drug sat along a packed dirt road leading family jalopies out into the scorching Badlands beyond town.

Given Wall Drug's outsized reputation as one of the most famous roadside attractions in America, I'm surprised to arrive in Wall and find it little more than a huddle of modest shops along a short but wide main street. Railroad tracks run nearby, while five looming grain silos dwarf the whistle stop. Quite out of scale with the town, Wall Drug encompasses 76,000 square feet and hosts more than two million visitors each year. When summer comes, two hundred people—almost a quarter of the town's population—will serve as Wall Drug employees during peak season. My meeting today is with Rick Hustead and his daughter Sarah. Rick is third generation, having come into the business through his father, Bill, and the store's founders, his grandparents, Ted and Dorothy.

There's nothing quite like stepping through the doors of Wall Drug. Inside what would otherwise be a cavernous space are many small, individual shops, each featuring a different theme. There is a tack store, T-shirt store, jewelry store, bookstore, some western-themed spaces, a cafeteria (complete with the free ice water advertised on the famous road signs), a narrow room dedicated to the history of the store, even a traveler's chapel. Each space has a distinct feel, and each is stuffed with merchandise—so much so that the crammed shops must be navigated carefully, and I often turn sideways to avoid bumping a rack of one item or a display of another. Every wall is plastered from floor to ceiling with something, with anything, with everything. While Wall Drug's keychains, coffee mugs, and stuffed animals are familiar to any road tripper with an appetite for kitsch, the labyrinthine vastness of this place is bewildering.

After wandering the maze of Wall Drug a while I happen upon a wall of jackalope shoulder mounts for sale. I find the display a little disconcerting, as it consists of sixteen jackalopes arranged in a long, unbroken row above a wide doorway. Only then does it occur to me how accustomed we are to seeing jackalope mounts one at a time, perhaps above a bar or pool table, and consequently I find this imposing phalanx of glassy-eyed antlered bunnies disquieting. While there is some variation evident in the antler shapes and the hue of the fur, the jackalopes look too much alike, as if a unique beast has been cloned, resulting in the stamping out of sixteen nearly identical twins. I've learned from studying Herrick family mounts that each jackalope maker has a unique style, so it seems likely that this sizable herd of little freaks has come from the hand of a single taxidermist.

Backing away from the packed wall of horned bunnies and related jackalopiana, I ask around for Rick and Sarah, who soon greet me and lead me upstairs into the old office space above the store. Rick is a lively guy, mid- to late-sixties, with grey hair, glasses, a black mustache and a friendly demeanor. His daughter, Sarah, is warm but quiet, with dark eyes, dark curly hair, and a trio of tattooed rings encircling one forearm. We chat about the Hustead family, the Depression-era origins of the store, its growth over time, and their vision for the future of this multigenerational business. Eventually I nudge our discussion toward jackalopes, for which Wall Drug has long been famous. Not only does the store sell a wide variety of jackalope mounts, T-shirts, postcards, and other such stuff, but they also have a large fiberglass jackalope statue on display in the courtyard behind the store. I ask Rick how long the iconic mega-jackalope has been part of the family's operation.

"Oh, around forty years, I guess," he replies. "Last summer a guy even proposed to his girlfriend on the big jackalope."

"Why do you think the jackalope has been so important to your business?" I ask, this time looking at Sarah.

"It's part of the mystique of this place. It's unique, and our store is unique, so it makes sense. Wall Drug and the jackalope are each an odd slice of Americana, and I think people enjoy that."

Rick nods in agreement. "Exactly, yeah, kind of a Wild West appeal too. The jackalope really fits at Wall Drug because it's so unusual. I think the energy we've created here even makes you want to believe that the jackalope is real," he adds, smiling enthusiastically. Rick is refreshingly earnest, like a grown man who isn't going to let go of Santa Claus, let alone give up on his family, town, or business.

I ask about the sourcing of the jackalope mounts I've seen for sale in the store, and the Husteads confirm that they do, in fact, have a single supplier. "We lay in as many as we can, because it can be rough some years," Rick observes. "There's not always enough fur—or, in other years, enough antlers—and we don't like to ever run out of our signature product."

Sarah nods. "Free ice water. Hand-painted road signs. Free coffee and doughnuts for vets. And jackalopes. That's what we're about. As fourth-generation I'm invested in the future of this place, but those are important links to our past. Our family history is entwined with the jackalope." I like Sarah. She listens more than she talks, she's direct when she has something to say, and she respects what her family has built while also bringing new vision to the future of that shared accomplishment. She's clearly the bridge between the past and future of the enterprise. For example, Sarah serves as the social media specialist for the business, work that has included a digital makeover of the old tradition of photographing oneself holding a handmade sign that reads "X Miles to Wall Drug." Such signs have been shared from every corner of the world, from the Taj Mahal (which, in case you're wondering, is 10,728 miles from Wall Drug), from the Paris Métro, even from Antarctica. I promise Sarah that when I return home to the western Great Basin Desert I will include this note in my jackalope book: "1,243 miles to Wall Drug."

I ask the Husteads if they know when jackalopes first appeared for sale in Wall Drug. I point out that while their family's mercantile opened in 1931, the first hoax jackalope was fabricated in the 1930s—perhaps as early as 1932—just 250 miles southwest of here in Douglas, Wyoming. Although the infamous rabbit and the store that is renowned for selling it are nearly coeval, Rick and Sarah, surprisingly, are unaware of the jackalope's canonical origin story. They do, however, confirm that their current supplier bought the taxidermy business from Jim Herrick, and they agree with me that it seems likely that the older Herricks may have supplied Wall Drug with mounts in the early years, though they can't be sure. They also think the *New York Times* got it right when, in Doug Herrick's obituary (published January 19, 2003), the paper reported that "Jim Herrick delivers four hundred jackalopes to Wall Drug in South Dakota three times a year, a small portion of his total production." The Husteads don't know what year the jackalope became a staple of the store's wares, though Rick adds that he can't remember a time when jackalopes were not for sale at Wall Drug. Because Rick's earliest recollections of the store in which he grew up include jackalopes, we can safely say that the world's most famous jackalope emporium has been selling antlered rabbits for well over half a century. But that only gets us back to the early 1960s. On my behalf Rick later checks with his ninety-year-old mother, Marjorie, who reports with certainty that jackalopes were already being sold at Wall Drug when she and her husband moved to town in 1951, when Rick was just one year old.

Although I've established that Wall Drug has been trading in jackalopes for at least seventy years, that still leaves some important questions unanswered. How and why did Wall Drug begin selling jackalopes? How many years before 1951 were they available? Was there an original supply line between Doug and Ralph Herrick and Rick Hustead's grandparents, who founded the store at about the same time the Herrick brothers fabricated their first hoax mount in Wyoming?

As I prepare to wrap up the interview, Rick calls downstairs and politely asks one of his employees to bring up a bag full of gifts for me: a history of Wall Drug, a Wall Drug cookbook, and a stack of various jackalope postcards, which are a staple of the endless supply of jackalope kitsch packing the shelves downstairs. Thanking the Husteads for their time, I ask my final question.

"Rick, why do you think people love jackalopes?"

He hesitates only a moment before replying. "They're super cute. They're bunnies. They really get a response. I mean, they're bunnies."

Now I look to Sarah, who knows exactly what she has to say about jackalopes. "I like them because I find mythical creatures fascinating," she replies, thoughtfully. "I'm always hoping they *are* real—that there's something magical hidden around the corner. A unicorn or a centaur, probably not. But a jackalope? That's something that *could* be real."

I shake hands with Rick and Sarah, thank them for their time and for the horned rabbit swag, and together we head back downstairs to the store. I snap a few pictures of father and daughter posing in front of a solid wall of jackalopes and then say a final goodbye. Before leaving the store, I purchase a grey T-shirt emblazoned with a logo reading "Jackalope University." This will be my fourteenth jackalope T-shirt, but I rationalize the purchase with a silent appeal to the old maxim that I should dress for the job I *want* to have. I decline the plastic bag and, shirt in hand, stop for a cup of free ice water on my way out. How many jackalope mounts have passed through Wall Drug and into the hands of tourists who have disseminated them, like weird seeds, across the country and around the world? There is simply no way to know. As for the postcards and other jackalopiana forever flying off the shelves here, it must add up to a number not only beyond reckoning, but beyond imagining.

Popular culture is saturated with jackalopes. To take sports as a thin but representative slice of jackalope pie, the Odessa Jackalopes, from Texas, are a team in the North American Hockey League. The Jammin' Jackalopes are the ultimate Frisbee team at Lebanon Valley College in Pennsylvania. The Jackalopes Mountain Biking Team rides out of Klamath Falls, Oregon. The Jackalopes are an amateur baseball club in Sioux Falls, South Dakota. A Houston, Texas volleyball team is called the Sand Jackalopes. A woman skating under the handle Jackalope is a roller derby star in Stockholm, Sweden. The Wild Jackalopes running club has its home in Denver, while Colorado is also home to the Jackalope rugby tournament. The Jackalope Skateboard Festival is held in Montreal, Canada. The Jackalope Triple Crown is a fencing competition hosted in Reno, Nevada. And, just in case you'd like to consider this a sport, Jumping Jackalope is a competitive axe-throwing venue in Spokane, Washington.

Unsurprisingly, the horned rabbit makes frequent appearances in association with alcohol. My own libation cabinet includes High West Bourye Whiskey, Horned Hare Bourbon, Harahorn Norwegian Gin, Drumshanbo Gunpowder Irish Gin, and Born and Bred Vodka, each of which features some form of jackalope on its label. Jackalope beers are legion. Especially notable are the craft beers concocted by Jackalope Brewing Company in Nashville, Tennessee—one of the few independent breweries in the nation owned by women—but you can catch horned rabbit brews all over the country. I recommend Gentleman Jackalope IPA from Mountain Forge Brewery in Broken Bow, Oklahoma, or, for the stouter of heart, the Jackalope Double IPA hopped up by Lamplighter Brewing Company in Cambridge, Massachusetts, just a mile down Broadway from Harvard. If you're seated on the grape side of the aisle, try the vintages released by Jackalope Wine Cellars near Salem, Oregon. If you aren't a drinker, you might be interested in the strain of cannabis called Jackalope, a cross between the Chocolope and Black Jack strains, which is said to taste of berries, sweet chocolate, and vanilla bean. If neither booze nor weed appeals, look for the running jackalope on the can of Hopscotch Dry Hopped Ginger Ale or relax with a decaf cappuccino at Jackalope Coffee and Tea House in the Bridgeport neighborhood on Chicago's South Side.

If you're looking for a jackalope restaurant or bar, almost any town will do, as the proliferation of burger flipping and booze slinging outfits named after the horned rabbit is astonishing. My favorites include the Jackalope Grill in Bend, Oregon, Jackalope's Bar and Grill inside the Tenaya Lodge in Yosemite National Park, and the Jackalope, one of the country's classic dive bars, on Dirty Sixth Street in Austin, Texas. Still on my to-do list are Jackalope Jack's Restaurant and Bar in Charlotte, North Carolina, the Crafty Jackalope in Bridgeville, Pennsylvania, and—this one purely aspirational—Bar Jackalope, an eighteen-seat, hidden-door backroom whiskey sipping library within Seven Grand bar in Los Angeles.

What else is named for the alluring mythical bunny? Easier to ask what isn't. The horned rabbit has its name on geographical features including springs, mountains, flats, ridges, and even a gas hydrates system in the Gulf of Mexico. (I myself narrowly failed in an attempt to legally establish the name Jackalope Bay for the desiccated arm of a remote high desert playa in northwestern Nevada.) The jackalope has also given its name to a summer camp in Washington State, a motel in Las Vegas, Nevada, a volunteer group in San Diego, California, a real estate firm in Park City, Utah, a pressure washing company in Ocala, Florida, a farmstead business in Roundup, Montana, a folk art store in Santa Fe, New Mexico, a photography shop in Annapolis, Maryland, a bakehouse in Boston, Virginia, and a dance recital video production company in Mesa, Arizona. Jackalope Customs and Accessories is an auto parts shop in North Texas. Jackalope Arts is an indie craft fair in Pasadena, California. Jackalope Jump is a bounce house rental outfit in Helena, Montana. Jackalope Beauty Lounge is a salon in Keller, Texas. Jackalope Hydroseeding is a landscape reclamation company out of Shelly, Idaho. Jackalope Meadows is an apiary in Sarasota, Florida. The Crafty Jackalope is a knitting shop in Vancouver, British Columbia. I'm pleased to report that Chicago's Jackalope Theater is now in its thirteenth season.

The jackalope is the school mascot for Rudolfo Anaya Elementary School in Albuquerque, New Mexico. The Continental Sausage Company in Denver, Colorado has pioneered jackalope sausage—a combination of antelope and rabbit—while Lunchables snack foods commercials feature cartoon avatars of the mythical jackalope along with the actual platypus. President Ronald Reagan displayed a buck-and-doe double jackalope shoulder mount in his Rancho Del Cielo home in Goleta, California. In a photo taken on November 25, 1987, at the ranch, Reagan uses his jackalopes to try to pull one over on National Security Advisor Colin Powell, whose laughter shows that he knows better. The Jackalope Freedom Festival, a gathering of anarchists held in the forest each summer, is said to be a community "where spontaneous order happens naturally, without any force, coercion, or aggression." Now that the Swedish roller derby, the Ronald

Reagan Presidential Library, and the anarchists have all appropriated the jackalope, there may be nothing left that the horned rabbit is unable to represent.

Within popular culture, the jackalope is present in every form of media imaginable, including music, drawing, painting, sculpture, literature, comic books, erotica, video games, television, and movies.

My favorite case study is music, which is among the most diverse and fascinating cultural expressions of jackalope appreciation, even fanaticism. There is Jackalope Records, an indie label in Sonoma County, California, whose logo, a clever homage to the old dog-and-gramophone recording trademark inspired by the painting *His Master's Voice*, depicts a horned rabbit listening intently at the sound of an old Victrola. The cover of swamp rock band Little Feat's *The Last Record Album*, created by noted countercultural comics artist Neon Park, depicts a jackalope in a Sonora or Mojave Desert setting that is, unaccountably, located in Hollywood. Rock icon Dave Grohl (originally of the band Nirvana) pulled off a five-minute comic video, *Super Saturday Night*, in which the members of his band, the Foo Fighters, play the roles of hapless players on the (fictional) San Fernando Jackalopes football team. The squad is so bad that the team's mascot, which is of course a jackalope, screams at the players ("You fuckin' suck! You guys suck! Goddamit! . . . You guys suck ass!") until Grohl runs full speed and violently tackles the mascot. Even British rock superstar David Bowie was a jackalope enthusiast. Bowie called his Fall 2002 concert run through Europe the "In Search of Jackalopes Tour," and he even kept a jackalope mount onstage near the drum kit behind him for that tour until it mysteriously vanished after the September 27 show in Bonn, Germany. Bowie's guitar tech, Andrew Burns, had an emergency replacement jackalope flown in from Wyoming to preside over the rest of the tour.

Jackalope songs span genres, from kids' music, folk, country, blues, and Americana to jazz, pop, rock, metal, and electronica. Songs named for the horned rabbit have been recorded by Colonel Claypool's Bucket of Bernie Brains, Luna Rose, the Gary Burton Quartet, Retro the Rift God, Jack Funk, Elektrodrei, Ashtongue, the Post Mortems, Hand Jam Tribe, Lovgret, Bunny Sings Wolf, the Fred Hersch Trio, GrandBob, Twink, Sir Orso Bruno, Fillmore, Yesterday & Tomorrow, Tear Up the Planks, and Silent Kol, to name just a few. There's the spacey, hallucinogenic "Shrooms and Jackalopes" by Arjun, "Day of the Jackalope" by hard rock band Clutch, smooth jazzy "Jackalope Utopia" by Dick Arnold, "Return of the Jackalope" by German stoner blues rockers Plainride, and "Creepy Jackalope Eye" by cowpunk shredders the Supersuckers, whose front man Eddie Spaghetti screams out "In a jackalope space, on a jackalope high / I'm staring you down, creepy jackalope eye." Sacramento post-hardcore rockers Dance Gavin Dance give us "Chocolate Jackalope," which can cleanse the acoustic palate for an electrified "White Jackalope" by the Consolation, an Americana band consisting entirely of—wait for it—dudes who live in Copenhagen, Denmark. All or Nothing's relentless chord banging on "Mauled By Jackalopes" suggests that the band is more All than Nothing.

In "Jackalope Rising," Bay Area death rockers the Phantom Limbs offer a ten-minute cacophony of feedback and cymbal crashes intercut with sampled Orson Welles dialogue. "Rocky Mountain Jackalope" by Spare Rib & the Bluegrass Sauce, is a straight up pickin' party, while "The Running of the Jackalope" by Lawn Jockey is a respectable fiddle and mandolin tune. "Jackalope Grimace" is played by Spectral Scrubber, whose Spotify biography consists solely of the observation that "owning musical instruments is like having pets that are only alive when you rub their bellies." Minneapolis indie rockers the Goondas shot much of the video for their single "Jackalope Jesus" in a taxidermy shop, while "Dirty Jackalope" is a frenetic electronica track by a UK musician called Missqulater, whose Facebook page inexplicably features a man with a cat's head in place of his own. A similarly exhausting techno track is the Mandovibes's "Jolt Jackalope," which proves once again that decades of noble attempts to kill the drum machine have proven fruitless.

I find that, much like a screw-top bottle of Boone's Farm and a spray can of Cheez Whiz, Teenage Exorcist's punked-up "Wild Jackalope" pairs nicely with "Jackalope of All Trades," a bass-heavy thrash anthem by Avoid the Humanoids. Vacabou, a trip-hop duo from Mallorca, Spain, logs in with a contemplative, Partridge-Family-on-acid tune called "Dear Jackalope." In a baffling mash-up of cowboy ballad and beach guitar anthem, Gary Small & the Coyote Brothers warn tourists about the "Snaggle Tooth Jackalope." The Pyrate Queen offers the elixir of "Doc Davis's Jackalope Serum," which sounds like the soundtrack to a street puppet show, but is actually about the tragic mauling death of the singer's grandmother at the fangs of a horned rabbit. An entry in the intercultural category is Goblin's "La Furia Del Jackalope," a bleak death metal ballad shrieked out en Español. Don Bodin's "Ghost of the Jackalope" is tuned to the nostalgic innocence of an old-fashioned carousel, while Widowpaw's "The Night of the Jackalope," by contrast, would be perfect as the score of a horror film featuring a homicidal clown. Prominent Native American flautist R. Carlos Nakai's band Jackalope played in a cross-cultural genre he liked to call "synthacousticpunkarachiNavajazz." Austin, Texas, rockabilly outfit Jolie and the Jackalopes are a multigenerationally jackalopeified act, as lead singer Jolie Goodnight's mother, Kimmie Rhodes (who has played with Dave Matthews, Emmylou Harris, Patty Griffin, and Willie Nelson), headlined a band called Kimmie Rhodes & the Jackalope Brothers back in the 1980s.

Tall tales have it that jackalopes are excellent harmonizing vocalists, but whether or not the legend that the jackalope will sing *with* us is true, it is certainly the case that the horned rabbit has often been sung *about*. There's something for everyone in jackalope music, as you'll discover if you dip an ear into my eclectic, genre-spanning assemblage of jackalope-related songs, which may be found on Spotify as a publicly accessible playlist called *On the Trail of the Jackalope*.

Of the more than one hundred jackalope songs I've gathered, the best-known is "Jackalope" by Grammy Award–winning Minneapolis-based American bluegrass and roots music duo the Okee Dokee Brothers. The song appeared on their Grammy-nominated 2016 album *Saddle Up*, which featured songs written by bandmates Joe Mailander and Justin Lansing during a month-long horse-packing trip along the Continental Divide. When I spoke with Joe about what attracted the Okee Dokee Brothers to the horned rabbit, he offered two replies. In a practical sense, he said, the jackalope was a good choice for his band, whose core audience consists of children and families. As Joe pointed out, "characters or animals with crazy attributes engage kids and help tell a story. And this one was an obvious choice for us because it plays into the myths of the West." But then Joe became more philosophical, commenting that a subliminal theme of *Saddle Up* is that "greater truths can come from stories that might not be factual. Stories, even though they're made up, can reveal truths that mere facts can't. It's ok to believe in something absurd," Joe said. "That's why the jackalope spoke to us. There's got to be some magic out there, right?"

The Okee Dokee Brothers's fans certainly love "Jackalope," which is among the most requested songs at the many live shows they perform around the country each year. In explaining the genesis of the song, Joe told me that he and his writing partner thought of "Jackalope" as alluding to a sort of spiritual journey—one that "can lead us to meaningful places" because it requires us to have faith in something beyond our everyday experience. When I closed our conversation by asking Joe why he thinks people love jackalopes—and why they love his jackalope song—his nuanced and thoughtful reply focused on the horned rabbit's hybridity. "We're all drawn to combinations of things. We love to see mixing and matching. In a way, it's a symbol of diversity, of bringing in different influences and mashing them together and seeing what they make. And isn't that a cool symbol for a nation of immigrants? Lots of different cultures coming together and seeing what we can make if we accept each other's quirky attributes."

Before we end our call, Joe says he has one last story he would like to share. He tells me that his three-year-old son has a jackalope mount (named, you guessed it, "Jack") on his bedroom wall, and that the boy loves it so much that each night he pets the animal on the neck before toddling off to dreamland.

As with musical genres, the jackalope appears in a vast range of visual arts media, from pictorial, graphic, and sculptural artworks to jewelry and tattoos. Robin Arthur's stylized jackalope art print depicts a horned rabbit whose skeleton may be seen through its body, whose eyes appear to consist of spinning sunflowers, and whose ears are festooned with stars, in a kind of Dia de Los Muertos–meets–Grateful Dead aesthetic. Amy Titus's striking painting *The Jackalope* depicts a magical forest rabbit whose horns do not branch like antlers but instead curve gracefully like the horns of a gerenuk, if only the gerenuk's horns also spiraled fantastically like those of an eland or springbok. Etsy sells a print resembling a page from an anatomy textbook, the title of which is also its caption: "Figure 1. Vasculature of the common jackalope, *Lepus cerastes*." Melbourne artist Emily Floyd's twenty-three-foot-tall gleaming black modernist statue of a horned rabbit greets guests at the entrance to the five-star Jackalope Hotel on Australia's Mornington Peninsula. Tarna the Jackalope Art Car is a mobile sculpture, a mutant vehicle welded together for moving display at the Burning Man Festival on the Black Rock playa in northern Nevada.

When Denver artist Bill Nelson contacted me to consult about jackalopes, I connected him with Mike Herrick, who made a custom jackalope using the techniques his father would have used. With an authentic Herrick jackalope in hand, Bill completed a striking piece of assemblage art consisting of the old-school jackalope mount nestled amid a field of dried wildflowers beautifully arranged within a gilded Victorian-style frame. James G. Mundie's woodcut *Lesser Horned Coney* is a meditative piece of observational art in the style of Albrecht Dürer's famous 1502 watercolor *Young Hare*, though this bunny has antlers so delicate and believable that the piece resembles a page from a Renaissance natural history folio. A more humorous effort along the same lines is an impressively authentic-looking actual Dürer, expertly photoshopped by the artist Madame Memento to include the horns of a pronghorn antelope. Rob Ozborne headed in a very different direction, creating a striking series of prints in the style of classic

Japanese monster movie posters in which a giant fire-breathing creature called JACKALOPE-ZILLA threatens to attack a city.

In "Jackalopes of the World," artist Lyndsey Green has imagined how different subspecies of jackalope might have evolved in different environments. Florida artist Andrew Williams offers a witty adaptation of the iconic Barack Obama "HOPE" image, now transformed to include a horned rabbit and the message "LOPE." Hannah Yata's jackalope-inspired mythical being, which she calls Dahlia, is an otherworldly rabbit with beautifully curving horns as long as its body. Brazilian tattoo artist Taiom has created body art of a jackalope so large that it fully occupies a man's chest, looking straight at the viewer, casually resting its paw on the edge of the frame in which its image is contained. My friend, the Nevada artist Reena Spansail, has done an inspiring watercolor, *Faster than the Speed of Light*, which depicts a captivating cosmic jackalope in full flight through a sidereal field of spinning stars. For years this beautiful piece has graced the sky-blue wall above my writing desk.

Among the most intriguing and gifted of horned rabbit visual artists is ceramist and decorative painter Kari Serrao, much of whose work consists of encaustic paintings, which are made using heated pigmented beeswax. In her series *Whimsical Woodland*, Serrao has depicted bears, foxes, raccoons, raptors, and canids with antlers—though the antlered rabbit paintings, of which there are more than fifty, are most prominent in the collection. A representative example of the series, *Morning Music*, is a large 36" x 48" encaustic painting depicting a whimsical horned rabbit in upright sitting position with beautiful, branching antlers and huge pink-cupped furry ears. The rabbit's long, whiskered face is a pastiche of greys, with a bright orange-and-black eye gazing directly at us. It is dressed in an ornate embroidered cape and high frilly Elizabethan ruff. Three robins encircle its ears and antlers.

Kari Serrao grew up in the Caribbean and lives in Toronto, but was painting in the south of France when I contacted her to discuss her work. When I asked about the source of her fascination with antlered rabbits, she explained that she had painted a mural in France that included a rabbit in an Elizabethan collar, and that while working on a subsequent commissioned decorative painting in Montreal, the client had asked her to put antlers on the rabbit rather than the deer. Since then she has been irresistibly drawn to this figure. "Rabbits and hares are especially appealing," she told me. "Compositionally, they have the added benefit of their ears; I love the relationship their ears appear to have with the antlers. I also find their eyes fascinating and they just seem to have a mystical quality about them." When I asked about her unusual choice to anthropomorphize the horned hares by representing them wearing ornate, lovely Elizabethan attire, Kari explained that "many of the animals in the series are considered to be pests by humans. By clothing them in such an aristocratic manner I was elevating their status, or elevating how humans may see them." And while Kari's fascination with the antlered rabbit did not emerge specifically from the folk tradition of the jackalope, that moniker did catch up with her work. Although she was only peripherally aware of the jackalope when she started painting the series, she now says "the portraits are constantly being referred to as 'jackalopes' when I show them."

The number of writers who have featured the jackalope in their work is impressive, and jackalope lit spans genres from children's books and comic books, to short stories and novels, to crime, sci-fi, gothic, fantasy, lyric poetry, and even erotica.

Stacey Balkun's poetry collection *Jackalope-Girl Learns to Speak* features a young female protagonist who, by imagining herself as a jackalope, seems to acknowledge her liminal status as a figure caught between childhood and adulthood. *Project Jackalope*, a young adult novel by Emily Acton, is an engaging whodunit that combines science, mystery, and ingenuity to solve a jackalope-related mystery. Another mystery, Ann Charles's *The Great Jackalope Stampede*, is the third book in her Jackrabbit Junction humor mystery series. Jackalopes appear in the Image Comics series *Proof.* A sly, politically-progressive jackalope—a sort of countercultural Bugs Bunny—is the hero of *Junior Jackalope* and *Tales of the Jackalope*, two 1980s comic book series by the gifted northern California cartoonist R. L. Crabb, who told me his inspiration came from a jackalope postcard he received during the 1970s.

Jackalopes are legion in children's books. Among my favorites, in part because of their charming and whimsical illustrations, are Julian Lund's *Have you Ever Seen a Jackalope?*, Baba Roy's *Song of the Jackalope*, Jonathan Jones II's *The Hope of a Jackalope*, Cyndera Quackenbush's *The Radioactive Rabbit*, Janet Stevens and Susan Stevens Crummel's *Jackalope*, and, best of all, *Juan and the Jackalope*, a story in verse by Rudolfo Anaya, the distinguished Chicano writer and author of *Bless Me, Ultima*, who died in 2020.

The horned rabbit has often been featured in the fantasy genre. In *The Titan's Curse* by bestselling young adult novelist Rick Riordan, the goddess Artemis sometimes transforms boys into jackalopes. The novel *Jackalopes and Woofen-Poofs*, the fifth book in Angel Martinez's *Offbeat Crimes* series, includes jackalopes along with paranormal cops and charismatic vampires. In Denise Low's novel *Jackalope*, the shape-shifting title character—who is a gender-bending Native American trickster figure named Jack/Jaq—encounters other trickster-cryptids including jayhawks, chupacabra, and aliens from Roswell, New Mexico. In Sean O'Brien's novella *The Jackalope Saves the World*, the horned rabbit is even figured as a superhero! In a darker turn, *Unnatural Tales of the Jackalope*, edited by John Palisano, is a multiauthor anthology of jackalope short fiction that tends toward the gothic/horror genre. For example, Kristi Petersen Schoonover's disturbing short story "The Thing Inside" concludes violently when a pregnant woman, tormented by a vicious pack of jackalopes, stabs at one that has crawled onto her, only to discover, to her horror, that the jackalope is purely imagined, and she has actually plunged the knife deep into her own belly.

Many jackalope stories imagine magical relationships between the horned rabbit and humans. For example, prolific Canadian novelist Charles de Lint's *Medicine Road* tells the story of a red dog and a jackalope who are magically changed by Coyote Woman into human beings. Ursula Vernon's "Jackalope Wives" (written under the name T. Kingfisher), which won the 2014 Nebula Award for science fiction and fantasy writing, tells the harrowing story of what happens when a man's plan to marry a jackalope goes terribly wrong, trapping the animal between human and nonhuman realms. Perhaps the inevitable conclusion of this trajectory of fantastic literary representations of jackalopes is Hannah Wilde's 2014 pornographic novella *The Jackalope Farm*, a work of "paranormal erotica" in Wilde's aptly named *Violated by Monsters* series. Jackalope porn, as it turns out, is every bit as funny as you might expect. In *The Jackalope Farm*, our first-person narrator, Kara, is hired by a not-at-all credible captive breeding program to harvest semen—in the most direct ways imaginable—from rare man-size horned rabbits. The sordid tale focuses less on the furry animals' horns than what is apparently a much larger part of their anatomy. In a delightful moment of comical understatement, Wilde's protagonist confesses that "the jackalopes are incredibly competent lovers."

The horned rabbit has also been spotted in video games, on television, and in films, cultural environments in which its appropriation is akin to what would once have been considered popular folk culture.

The jackalope appears in early video games such as *Redneck Rampage Rides Again* (1998), which takes place on a jackalope farm. In part of this first-person shooter game, players can stalk their horned prey to the tune of "Dueling Banjos" and blast the poor bunny to smithereens while drawling that "them jackalope innards is just as good the second day." In *Urbz: Sims in the City* (2004) players can adopt a jackalope as a pet, while in *The Sims 2* (2004) it is possible to purchase jackalope mounts. The horned rabbit is also featured in the wildly successful *Red Dead Redemption* (2010) in which it is necessary to kill a jackalope in order to unlock the so-called Expert Hunter Outfit. In *Guild Wars 2* (2012) a player can pursue the elusive mythological creature in the Osenfold Shear of the Shiverpeak Mountains, or even use the Potion of Jackalope Transformation to shape shift into a horned rabbit. *Far Cry 5* (2018) extended and deepened the possibilities by including the jackalope not only within the game itself but also in a fascinating dream sequence in which the player witnesses a lovely horned rabbit nibbling plants in a glistening, mist-shrouded, emerald meadow. In *Hogwarts Mystery* (2018), which takes place within the Harry Potter universe, students at Hogwarts School of Witchcraft and Wizardry learn about the jackalope in Professor Silvanus Kettleburn's Care of Magical Creatures class, where they are informed that the animal's antlers were used as wand cores by seventeenth-century wand makers. The creators of the MMO (massively multiplayer online game) *Feral* (2020) describe it as an "online world where you'll become your own creature of myth." Among the options, players may choose to customize a jackalope avatar for game play.

Television producers seem to love jackalope mounts, which may be spotted in programs from web series to cartoons to sitcoms. Among my favorites, a T-shirt worn by Charlie in the FX comedy series *It's Always Sunny in Philadelphia* depicts a drunken leprechaun in a violent fistfight with a jackalope. During the 1990s the horned rabbit played a prominent role in a number of shows. Swifty Buckhorn, a horned rabbit, starred in *Wild West C.O.W.-Boys of Moo Mesa*, a Saturday morning ABC animated series by comic book artist Ryan Brown, best known for his work on Teenage Mutant Ninja Turtles. A jackalope named Jack Ching Bada-Bing often pranked people on the ABC series *America's Funniest People*. Even a mainstream sitcom like *Frasier* featured our horned hero. In the episode "A Tsar is Born," Martin Crane's jackalope mount prompts a mock-heroic quip from his well-heeled son, Niles, who condescendingly refers to the jackalope as "Texas's answer to the minotaur."

Mutant rabbits appeared on the big screen long before the cinematic debut of their horned cousins. In my favorite, the 1972 cult classic *Night of the Lepus*, a hormone devised to control an exploding rabbit population goes awry, turning the bunnies into human-size flesh-eating monsters (tagline: "Your hare will stand on end!"). A few years later the Monty Python troupe offered a hilarious take on the cutthroat rabbit with the unforgettable Killer Rabbit of Caerbannog in *Monty Python and the Holy Grail*. These deadly precursors opened the way to horned rabbits in many cinematic forms. The horned rabbit is the star of several short films, including Joseph Guerrieri's *The Jackalope* and Daryl Della's *After the Jack*, in which the main character, a cowboy named Red, thinks his path to glory as a roper requires that he capture a jackalope. The best of these shorts is *Boundin'*, created by Pixar Animation Studios in 2003. In this charming five-minute animated story set in the American desert, a timid sheep who is devastated after having been captured and shorn by a heartless unseen human is consoled by a bouncing horned rabbit who is described by the film's song as "this sage of the sage, this rare hare of hope." Having taught the sheep a comforting lesson in self-acceptance and resilience, the heroic animal bounds off, as the final words of the song suggest the horned rabbit has always functioned as an avatar of hope: "Now in this world of ups and downs / So nice to know there are jackalopes around."

While jackalope shoulder mounts may be spotted in a number of films—including Rob McKittrick's *Waiting* and Ang Lee's masterpiece, *Brokeback Mountain* (both released in 2005)—films featuring the jackalope as a central plot device exist in several cinematic genres. In animated kids' movies, the horned rabbit drives important parts of the action in *Scooby Doo and the Alien Invaders*. In foreign language pictures, Montreal filmmaker Jean Fugazza's *Melissa et le Jackalope* tells the story of a young Quebecer whose love for her rabbits prompts her to transform them into jackalopes. Karl Shefelman's film, *Looking for the Jackalope*, features the horned rabbit as a guiding spirit—a personification of central protagonist Jordan Sterling's past, which he is on a troubled quest to rediscover.

Unsurprisingly, film comedies have often deployed the jackalope. Working in the mockumentary, a genre dear to me, director Dustin Carpenter has great fun with *Stagbunny*, a would-be documentary in which cryptozoologist Garvis Thurston's search for the jackalope leads him to awkward, entertaining interviews with Wyoming locals, hunting guides, eyewitnesses, and even a ghost tracker. However, the king of jackalope film comedy is the inimitable Dave R. Watkins, a wild indie filmmaker from north Georgia whose 2005 horror comedy *Curse of the Jackalope* is as self-aware as any horror parody could be. But the deliberate badness of the film was just the beginning, because the best-worst was yet to come. Even while making *Curse of the Jackalope*, Watkins and his codirector, Michael Friedman, had already planned their next move. The following year they doubled down on their self-ironizing approach by producing *Return of the Jackalope*, an entertaining mockumentary about an attempt to reunite the cast and crew of the awful *Curse of the Jackalope*, now lovingly referred to as "the lowest grossing film of all time."

When I got Dave Watkins on the phone, I was struck by the lilt and cadence of his southern drawl. For a guy who made movies about homicidal jackalopes, he was thoughtful, even quiet. Dave told me he had studied video production at the Art Institute of Atlanta and worked in the film industry but found he was happier working a day job at his family's lumber business and devoting his free time to his own creative work, rather than being paid to labor on other people's projects. "I'd rather do the creative part than the logistical part," he told me, noting that most of his industry gigs had been as an assistant director. A prolific filmmaker who has produced dozens of shorts and four web TV series with a total of more than one hundred episodes, Dave is attracted to the horror genre but finds producing what he calls "straight horror" intense and exhausting. "I enjoy making horror movies, but it's much more fun to do comedy. With horror the energy is more drawn-in and dark, and you start to feel the weight of it after a while. Comedy is more of a release. Even though you're tired at the end of the day with comedy, it feels more upbeat than when you have to get in the darker place with horror."

Dave told me his jackalope idea came from a mount that happened to be on the wall of one of the homes in which he was shooting another film. I wondered if that mount might have been sold at Wall Drug, or fabricated by Frank English, or Mike Herrick, or Mike's cousin Jim in Douglas, Wyoming. That jackalope became a kind of mascot for Dave's crew, so when the mockumentary/horror/comedy mash-up idea was born, they decided that a murderous jackalope was just weird and funny enough to be their monster of choice. Like any legit documentary about the making of a film, the *Return of the Jackalope* mockumentary is peppered with clips from *Curse of the Jackalope*—clips so exquisitely terrible as to make clear to viewers that the idea of a cast reunion is a serious mistake. In one scene, the killer jackalope (played in a preposterous, homemade costume) murders a girl scout in a moment so gratuitously bloody as to prompt laughter rather than fear; in another, a man about to be slain by the giant jackalope actually offers the killer rabbit a cold beer in an (unsuccessful) attempt to mollify him. The clips, in their irresistible awfulness, support the film's comic proposition that a jackalope horror film so dreadful that it should never have been made is certainly not a choice candidate for revival.

Dave told me the wonderful story of how he and his codirector took clips from the intentionally over-the-top *Curse of the Jackalope* to the Dragon Con pop culture conference in Atlanta, where they persuaded a number of excellent professional film actors, including Bruce Hopkins (*The Lord of the Rings* franchise), Ernie Hudson (the *Ghostbusters* franchise), and Garrett Wang (the *Star Trek* franchise), to watch the clips and then offer spontaneous, tongue-in-cheek commentary on the film—commentary they later edited into their mockumentary *Return of the Jackalope*. At one point Hudson, playing it so straight that the moment is hilarious, comments that "I love Disney movies, and I've done my share of nice movies. But it's also nice to see people get their heads chopped off sometimes. The blood is like detergent. When it's all over, you sit there and you feel washed over." And Garrett Wang really gets into the spirit of the project when, with intense mock seriousness, he testifies that *Curse of the Jackalope* "really impacted my life because I realized that it's so bad that it's actually genius."

Like a jackalope mount, tall tale, or postcard, Dave Watkins's mockumentary appeals not despite its failure to fool us, but because of it. Like so many of the musicians, painters, cartoonists, animators, website and video game designers, and television producers who have also featured the horned rabbit in their art, Dave understands that the jackalope can play almost any role, from adorable bunny to savage murderer. In this sense his film embodies the same playfulness—and also the same lively tension between innocence and viciousness—that has long been a staple of traditional jackalope folklore.

Chapter 6
Necessary Monsters

It can be difficult to determine what a fabulous being is.

—Malcolm South,
Mythic and Fabulous Creatures

I love the work of Argentinian writer Jorge Luis Borges for its wonderful exploration of the limits of the actual. In *The Book of Imaginary Beings*, Borges describes a menagerie of fabulous creatures including the golem, roc, hydra, minotaur, siren, and unicorn. Also cataloged is the Six-Legged Antelope of Siberian myth. The divine huntsman Tunk-poj pursued this antelope all over heaven before finally capturing it. However, the animal's extraordinary speed made plain that mortals would never be able to hunt it successfully, and so it was decided that antelopes would from that day forward be quadrupeds. Borges has included the Garuda, an animal with the head, wings, and talons of an eagle, and the chest and legs of a man. The Hindu god Vishnu is often depicted riding on this hybrid bird-man, who is so learned as to teach benighted humans about the genesis of the universe. There are many other mythological fusions in Borges's wonderful bestiary: the Chimera, which in Homer and Hesiod is an amalgamation of lion, goat, and serpent; and the gryphon, the ferocious winged monster of Herodotus and Pliny, whose talons are so immense that men make drinking vessels of them.

Into his literary tapestry of imaginary creatures, Borges has woven the panther and remora, which actually do exist. Even the kraken, that fiercest of mythological sea monsters, may simply be what millennia of sailors' yarns have made of massive but rarely encountered cephalopod species such as the fifty-foot-long Giant Squid (*Architeuthis dux*) or the Colossal Squid (*Mesonychoteuthis hamiltoni*), which, at more than 1,000 pounds, is the largest invertebrate on the planet. Borges has also included among his imaginary beasts the pelican. "The Pelican of everyday zoology is a water bird with a wingspan of some six feet and a very long bill whose lower mandible distends to form a pouch for holding fish," he writes about this improbable bird, while "the Pelican of fable is smaller and its bill is accordingly shorter and sharper." Are we to conclude that the actual pelican is more fabulous than the mythical pelican? Borges's assemblage of fantastic beings has the odd effect of obscuring rather than delineating the permeable boundary separating the real from the imaginary, reminding us that many beasts of fable join the kraken in having connections to real-world species both extinct and living.

I have read Borges's *The Book of Imaginary Beings* from Abtu and Anet, the sacred fish of the Egyptian sun god Ra, to the Japanese Eight-Forked Serpent of Koshi, to the Persian Manticore, which has the body of a lion, the face of a man, and the tail of a scorpion—and all the way down to Zaratan, a sea beast so massive that sailors in each of the seven seas have, at their peril, mistaken its broad back for an island. "We are as ignorant of the meaning of the dragon as we are of the meaning of the universe, but there is something in the dragon's image that appeals to the human imagination," wrote Borges. "It is, so to speak, a necessary monster." So, too, the jackalope, a necessary monster which is the chimera or gryphon of these windswept expanses of western desert, mountain, and prairie. The horned hare inhabits an enchanted land which, when engaged by the wildness of the human imagination, renders the magical incontestably real and the real inexplicably magical.

Despite Borges's richly populated imaginative universe, we live in a sometimes terrifyingly real world in which climate change, habitat loss, and myriad other environmental pressures have resulted in a devastating loss of biodiversity. It is thus a wonderful surprise that so much of the living world remains undiscovered, and that so many new species have been found during the past century—even as we have radically reduced the earth's capacity to sustain a rich diversity of flora and fauna. Fewer than two million species are known to science, which means that at least 80 percent of species on our home planet remain unknown. I find it inspiring that each year scientists discover around 18,000 new species. For example, during the first half of 2020 alone, researchers at the Biodiversity Unit at the University of Turku, in Finland, described seventeen new spider species, twenty-three new insect species, and a new monitor lizard, and these previously unknown creatures came from around the world: the Middle East, India, the Amazon, Europe, and the Pacific islands.

Animals discovered during the past century are certainly not limited to spiders and insects. Not recognized as a species until 1933, the highly intelligent Bonobo, a great ape native to the Congo Basin, shares 98.7 percent of its DNA with humans. The ten-foot-long Philippine Crocodile, a freshwater animal, was discovered a few years later in 1935. When the Coelacanth—a massive, lobe-finned fish often described as a "living fossil"—was caught by a fisherman in 1938, it was identified as a member of a group of fishes thought to have been extinct for sixty-six million years. The Javelin Spookfish, discovered in 1958 off British Columbia, has six eyes that allow it to see at the inky depth of 3,000 feet. The Megamouth Shark, which grows to seventeen feet long and is among the largest species of living shark, was first captured off Oahu in 1976. The year 1978 saw the discovery of one of the most poisonous animals on our planet: the spectacularly beautiful Golden Poison Frog, whose remarkable batrachotoxin secretions are used by the indigenous Choco Emberá people of the Panamanian rainforest to tip their hunting arrows.

Many dramatic discoveries have been even more recent and seem to me every bit as incredible as the imaginary animals of Borges's bestiary. Southeast Asia's Giant Freshwater Stingray, which weighs a half ton and has a body disc six feet wide, wasn't described until 1990. In 1992 the Saola, a large, antelope-like ox thought to have been extinct for four million years, was found in the Vu Quang Nature Reserve in Vietnam. The Black-crowned Dwarf Marmoset, a six-inch-long monkey, was discovered in the Brazilian jungle in 1996. The Borneo river shark was rediscovered in 1997 after having been known only from a single museum specimen dating to the 1890s. Found in Indonesian waters in 1998, the Mimic Octopus has the fantastic ability to avoid predators by disguising itself to resemble other creatures, from jellyfish and crabs to flounders and sea snakes. A species of microhylid frog from Papua New Guinea (*Paedophryne amanuensis*), located in 2012, is the world's smallest known vertebrate at less than a third of an inch in length. 2019 brought the discovery in the Galapagos Islands of a century-old female Fernandina giant tortoise (*Chelonoidis phantasticus*), a species last sighted in 1906. Also in 2019, divers discovered a new fish—a vibrant purple fairy wrasse—on a 200-foot-deep reef in eastern Zanzibar and named it *Cirrhilabrus wakanda* after the fictional African kingdom in the *Black Panther* series of comics and, more recently, Marvel universe movies. A new species of green pit viper discovered in Arunachal Pradesh in India in 2020 was named *Trimeresurus salazar* for Salazar Slytherin, one of the snakiest characters in J. K. Rowling's *Harry Potter* saga.

In a world so rich with startling diversity, so full of captivating strangeness, does it seem so outrageous that a horned rabbit might actually exist? These remarkable examples of fantastic and previously undiscovered creatures being brought from the realm of the imaginary or mythical into the world of the described and cataloged reminds us that the boundary between the known and unknown remains delightfully unstable, even into our own time.

Despite the liberating energy of Borges's wild imagination, it might be argued that an imaginary being is, by definition, not a being at all. And yet, all cultures include such beings as a part of their folklore and mythology, and sometimes depend on them for an understanding of the world—or, at least, for an appreciation of the things of this world that remain beyond our understanding. Mesopotamian culture is replete with fabulous beasts, including the half-scorpion, half-human figures that guard the gates of the sun god Shamash's sacred mountain in the *Epic of Gilgamesh*. The mythical beasts of ancient Egypt include the serpopards, with their slinky bodies, long necks, and leopard heads, and the gryphons, whose feline torsos sport wings and the head of a falcon. In book eight of his magnum opus, *Naturalis Historia* (circa 77 C.E.), the Roman natural philosopher Pliny the Elder included the unicorn, phoenix, manticore, and werewolf. Bestiaries, prominent in the medieval period, were vast compendia of marvelous animals whose cultural function was to supply allegorical meanings that assisted humans in organizing their moral universe. For example, while ancient pagan myths figure the unicorn as a one-horned animal that can only be tamed by a virgin, Medieval Christian scholars co-opted this narrative, refiguring the unicorn as an allegory for Christ in his relationship to the Virgin Mary.

Although European Renaissance naturalists came under the influence of early scientific methodologies, they continued to include many fantastic creatures in their works. Even into the eighteenth century, the great taxonomist Linnaeus recognized that his ambitious attempt to comprehensively catalog the living world would necessarily be limited by animals that remained beyond human ken. As W. Scott Poole writes in *Monsters in America*, "Linnaeus even left open the possibility that a category needed to be created for dragons, and his genera included a spot for troglodytes. Much like Voltaire, Linnaeus had little doubt that further exploration of the globe promised to yield a bounty of fantastic creatures with bizarre anatomies. Linnaeus perhaps represents the Enlightenment attitude toward the monster, the idea that the wondrous remained wonderful even as it became classifiable." Linnaeus seems to have intuited what Borges would later assert: in a world where the real is so often fantastic, it is plausible that the fantastic may sometimes prove to be real.

Many imaginary animals—including the garuda, chimera, manticore, griffin, mermaid, and, of course, the jackalope—are hybrid beings comprising features from different animals, or combining aspects of the human and animal. We have always been fascinated by hybridity, by sanctioned or illicit boundary crossing, by the possibility that the things of this world might miraculously appear in unforeseen, whimsical, or grotesque conglomerations. These composite beasts often function as trickster figures who rupture the boundary that otherwise separates the human from the nonhuman, the known world from the mystery that envelops it. According to Lewis Hyde, tricksters are the "lords of in-between" who function as intermediaries between worlds: "Trickster is a boundary crosser. Every group has its edge, its sense of in and out, and trickster is always there, at the gates of the city and the gates of life, making sure there is commerce. . . . Trickster is the mythic embodiment of ambiguity and ambivalence, doubleness and duplicity, contradiction and paradox."

Artist Lily Seika Jones's wonderful illustration of the jackalope which appears in *The Compendium of Magical Beasts* captures the spirit of the horned rabbit as an animal which inhabits the borderlands between the real and imaginary. The jackalope too is a trickster, a boundary crosser, an imagined creature whose hybridity signals its composite, magical nature, and also its boundary-violating playfulness. As Malcolm South observed, "Sometimes a writer's aim in placing fabulous creatures in the primary world is to satirize the inability of human beings to perceive the marvelous." So too the jackalope, a fabulous being placed into the real world not only to test us by straining the boundaries of the actual but also able to expand our sense of the possible and bring us joy. If a horned rabbit is less weighty than a half-ton freshwater stingray, it is, like that equally implausible animal, a fine reminder that the world is still larger than ourselves, and that the human imagination is the product of an evolutionary process that has graced us with an insatiable appetite for the incomprehensible and the marvelous.

Museums are places where we go to witness incredible things—things we might be inclined to deny the existence of were they not displayed in a public space defined by an unassailable cultural authority that certifies their authenticity. For example, I will never forget the moment I laid eyes on the giant squid (*Architeuthis kirkii*) at the American Museum of Natural History in Washington, D.C. At 500 pounds and twenty-five feet long, the specimen before me was undeniably real, and yet its physical presence only enhanced my disbelief. As the museum's website pointed out, "scientists still don't know how deeply giant squids dwell within the oceans, how many individuals there are, how many giant-squid species exist, how long they live, how fast they grow." The museum's curator, paleontologist Neil Landman, recognized that what we know about the world simply offers a salutary reminder of what we do not know. "In many ways," Landman observed, in reflecting on the massive cephalopod specimen, "the deep oceans are less well known than outer space."

Unlike the American Museum of Natural History, two of my favorite museums do not exist. Or do they?

The first of these is the online Museum of Hoaxes, curated by the historian of science and popular science writer Alex Boese. Established in 1997, this museum "explores deception, mischief, and misinformation throughout history, playing host to a variety of humbugs and hoodwinks—from ancient fakery all the way up to modern schemes, dupes, and dodges that circulate online." The museum's incredible range of materials spans from the Middle Ages through contemporary culture and covers fraudulent photos, April Fools' jests, forgeries, celebrity death scams, astrology, the paranormal, satirical art hoaxes, and even pareidolia, which is the tendency of humans to incorrectly perceive a pattern of meaning in a neutral object. (A well-known example is the Miami woman who saw the Virgin Mary in her grilled cheese sandwich; in 2004 the decade-old sandwich sold on eBay for $28,000).

This immensely entertaining and informative website includes the jackalope as part of a "Tall Tale Creature Gallery" alongside myriad other fabulous beasts, including the abbagoochie, antennalope, drop bear, fur-bearing trout, Pacific Northwest tree octopus, rackabore, snouter, tripodero, and woofen-poof. In his accompanying book, also called *The Museum of Hoaxes*, Boese features a few of his favorite tall-tale creatures, including "the upland trout (a fish that lives in trees), the treesqueak (another tree-living mammal, which can be heard squeaking whenever the wind blows), the snow snake (the bane of skiers), the gumberoo (which cannot be photographed because its image will explode), the squonk (a morose creature known to literally dissolve into tears), and the rubberado (whose flesh, if eaten, will cause one to bounce)."

I had wanted very much to visit the Museum of Hoaxes on my next trip to southern California, although the directions provided under the "Finding Us" tab on the museum's website raised my suspicions: "We're based in San Diego, California. If you're in the downtown area, get on I-5 north and keep driving until you see a giant floating jackalope off to your right. You can't miss it! If you reach LA, you've gone too far. Turn around and try again. Just remember that the giant floating jackalope will now be on your left-hand side." I soon abandoned the hope of ever reaching the museum, but nurtured a new hope that a phone call to Alex Boese might clear a few things up. I was anxious to learn more about the person who had devoted a mountain of time and energy to archiving and celebrating world-class contributions to the fine art of fooling people.

When I connected with Alex, who had recently relocated to Phoenix, I asked him about what I had come to understand as the absence of a physical museum located in San Diego—or anywhere else. "I still hear from school teachers who say, 'we're taking a field trip to San Diego, can you tell us more precisely where this is?'" He chuckles, adding quickly that he always lets people in on the joke and would feel badly if anyone actually went looking. "I thought it was funny to have a hoax about hoaxes. So the Museum of Hoaxes is itself a hoax. That's just the kind of silly thing that appeals to me," he told me.

I asked Alex how he became interested in hoaxes and, more specifically, in fantastic creatures. He explained that while working on a graduate degree in the History of Science at University of California, San Diego he had stumbled upon the Great Moon Hoax of 1835. "I have a kind of contrarian streak in me," he explained. "I was spending so much time in grad school, where you're expected to take stuff very seriously. The lack of seriousness with the hoaxes—that humorous element really appealed to me. It was so unlike what we were supposed to be writing about." He found himself asking, instead, how hoaxes might be considered part of a serious history of science, even persuading his mentors to allow him to write a doctoral dissertation on hoaxes. Before long he had created the website, which took off and soon led a publisher to invite Alex to write a book about hoaxes. He has since gone on to write a number of fascinating books of popular science, including *Elephants on Acid*, *Electrified Sheep*, and *Psychedelic Apes*.

I asked Alex how he conceives of the distinction between "hoax" in the pejorative sense and "hoax" in the benign or playful sense. "Tall tales may be where my real love is," he told me. "I got especially fascinated by the history of April Fools' Day, and that has its own little bestiary that comes along with it. All kinds of fantastic creatures are invented for April Fools' Day." April Fools' Day also charmed him because of the harmless nature of the deception it usually inspires. "The kind of hoaxes that aren't necessarily malicious—there's an element of play with them—those are always the ones that appeal most to me," he explained.

I was anxious to know why Alex had chosen the jackalope as the imaginary animal of choice to (not) lead people to his imaginary museum (a jackalope postcard also graces the cover of the paperback edition of his book). "Jackalopes are my favorite fantastic creature," he said, without hesitation. "I've got my own mounted jackalope. I've got a huge jackalope postcard collection. Any time I see any kind of jackalope thing—a little statue or whatever—I immediately buy it up. I've been doing that for twenty or thirty years now. I've accumulated a mini-museum of jackalope stuff." Speaking as a guy so obsessively committed to jackalopiana that my own family had to stage an intervention, I could easily relate to Alex's passion.

I ended the conversation with my fellow horned rabbit aficionado by asking my enduring question: "Alex, why do you think people love jackalopes?"

"Among the fantastic creatures, the jackalope is a charismatic character. Rabbits are inherently silly creatures, and then you put these antlers on them and you have this fighting warrior rabbit. That humorous, silly element manages to perfectly capture the animal." I appreciated Alex's emphasis on that contrast between comedy and badassery, and I was reminded of the Herrick jackalope I had seen in the Pioneer Museum in Douglas, Wyoming, which was at once hilarious and menacing.

"Also, it's not that hard to believe that it might be real," he continued. "Certain fantastic creatures are obviously fake, but when you show people a picture of a jackalope—and I've done this myself, a lot—they always ask, 'Is that real?' Maybe that's part of its appeal: something about it looks like it could be true, but when you realize that it's not true it's totally ridiculous, so it captures a perfect tension there."

The other nonexistent museum in which I take genuine delight is the Merrylin Cryptid Museum, which is curated by another Alex: London-based illustrator, writer, sculptor, and musician Alex CF. The museum's webpage explains the origin of the collection it exists to display:

> In 2006, a trust was set up to analyze and collate a huge number of wooden crates found sealed in the basement of a London townhouse that was due for demolition. Seemingly untouched since the 1940's, the crates contained over 5,000 specimens of flora and fauna, collected, dissected, and preserved. . . . But the most curious aspect of this discovery was the man responsible for its existence—the enigmatic, mysterious gentlemen that had gathered together a wealth of relics that challenged our understanding of nature; of species that had never been witnessed by

> the modern world, of objects which defied physical laws, Lord and Professor Thomas Theodore Merrylin.

There follows an impressively detailed narrative explaining how Merrylin, born in 1782, rose to become an accomplished naturalist, traveling the globe in search of unusual specimens—including many that had been dismissed or discredited by other scientists. The brilliant professor was also a cutting-edge theorist, postulating then-unimagined ideas such as the multiverse, time travel, and quantum mechanics, even as his work navigated the fringes of the dark arts and revealed a deep affinity for the occult.

I was two beers and three hours down this wonderful rabbit hole when I located a 2014 interview given by Alex CF to the horror magazine *Diabolique*. In it he was asked whether, before taking on the role of curator of the collection, he believed in cryptids—animals whose existence remains scientifically unsubstantiated. His reply speaks to the allure of Merrylin and his quest:

> I definitely had an interest in unknown things, but much of cryptozoology lacks the physical evidence that gives a species a natural history, a biology. I think the quality that attracted me to this work was Thomas. His diaries are filled with detailed analysis, annotated studies, and drawings of each of his finds. He stumbled upon unknowable things that speak of sciences and experiences we are yet to know, and species that seem to only exist within this collection. Yet there are the physical specimens that are his legacy. I have always been attracted to the idea that there are certain individuals who can look upon the world and see beyond.

I am obliged to reveal that Lord Merrylin is fictitious, as is every detail of his extravagant backstory, and that there exists no physical museum one might visit to view the amazing artifacts preserved in his collection. However, this elaborate narrative—the obsessed scientist, the rediscovered specimens, the dark mysteries of unknown sciences revealed—functions brilliantly as a framing tale within which Alex CF's own extraordinary artwork is displayed. I was far more captivated by CF's collection than I would have been by a real museum, which is, after all, necessarily limited to being a museum of the actual. For example, in the astonishing "Specimens" section of the museum's website, miracles and wonders abound. Here we find *Homomimus alatus* (Common Fae), a carnivorous winged biped resembling a small desiccated humanlike skeleton that has body-size, brown-and-white striped wings like those of a giant moth. It is, in effect, a faerie, one as lovely as it is grotesque. Also included is the *Homomimus aquaticus*, or Icthyosapien, a fishlike species which can "emit various human like vocalisations, which, purely by chance, can sound somewhat like a small child or woman singing. . . . This may be the source of the siren mythos." The specimen itself is a translucent, sinewy, lizard-like creature with dual dorsal fins and a face seen only in nightmares.

Other specimens are recognizable from folklore and mythology, such as *Homo Lupus* (Lycanthrope), a symbiotic hominid better known by its common name: werewolf. Alex CF's Lycanthrope head, preserved in what appears to be an authentic Victorian display case, is spectacular: as terrifying as any monster to emerge from a campfire story and as fascinating as any specimen displayed in a natural history museum. The detailed species description informs us that most Lycanthropes are born, but that on "rare occasions a human can become infected with Lycanthropy. The Lupus viral strain is a particularly hardy pathogen. It infects the host via saliva and blood and instigates cellular mutations. These mutations are rather unique as they not only alter the genetic make up of cells, it actively retrofits the host to suit its needs." In this way the Thomas Merrylin frame narrative creates a raison d'être for the physical artifacts, which are in turn described using the language of science, leaving the viewer with an irresistible sense that something that could not possibly exist looks and sounds as if it does. Viewing the collection's specimens gave me a feeling similar to the one I experience while gazing at a jackalope: it appears too fantastic to be real . . . and yet!

What other unusual creatures are included in Merrylin's collection? Easier to ask what is not included, as the wild imagination of Alex CF has inspired him to create both artifacts and text to accompany a wide range of outré species, from vampires, goblins, and gnomes to nymphs, dryads, and succubi. Among these wonders of nature we discover a fanciful quadrupedal herbivore called the Jackalope (*Lepus temperamentalus*), which is represented by a single small specimen lying gracefully in the fetal position. The animal has some fur remaining but, as is common with long-stored specimens, it is mostly a desiccated mass of wrinkled skin with its skeleton nearly protruding through the shrunken hide. Emerging from its head is a single gracefully branching antler. As with so many of Alex CF's creations, the creature is both horrible and beautiful. Like that giant squid I witnessed, it is a specimen whose death is redeemed by the fact that it brings marvelous new possibilities to life.

Alex Boese astutely observed that the jackalope perfectly embodies the tension between the authentic and the fantastic. Likewise, the genius of Alex CF's Merrylin Cryptid collection lies in its imperceptible blending of what might be real but is not with what cannot be real but appears to be. Like Borges, who curated a literary collection of imaginary beings, both of these curators of whimsical museums recognize the vital role of imagination in shaping our understanding of the natural world. While any zoologist can tell you that many animals once thought to be real are in fact imaginary, they must also acknowledge that many animals once presumed mythical turned out to be genuine. The platypus was once a cryptid, as was the giant squid, okapi, Komodo dragon, mountain gorilla, manatee, and many other wonderful creatures. As the guy who has the dubious distinction of being the world's most zealous jackalope researcher, I have a similar feeling about the prospects for my own idiosyncratic pursuit of the horned rabbit. I have tracked an imaginary being for so long and with such tenacity and passion that it has become real to me.

In the opening chapter of this book I shared the jackalope's canonical origin story: Douglas and Ralph Herrick created the first hoax jackalope mount near Douglas, Wyoming, sometime in the 1930s. But even as I believe that story to be true—after all, I made a one thousand-mile pilgrimage to Douglas specifically to explore it—I also referred to the jackalope birth narrative as "stubbornly unverifiable." Here is the main reason why that uncertainty is perennial: *horned rabbits actually exist in nature!* And if horned rabbits are real, how then can it be accurate to refer to the jackalope as imaginary?

We will look into this more deeply later, but for the moment a brief description of the actual horned rabbit will be helpful. The "horn" of the horned rabbit is a keratinous carcinoma, a growth that forms on the animal, usually on its head or face. These carcinomas can metastasize, or they can starve the rabbit by occluding its mouth and inhibiting its ability to feed. The malignant growths—which usually occur in cottontails but may also infect jackrabbits, European rabbits, snowshoe hares, and other leporids—are caused by a virus called Shope papillomavirus. While breakthrough scientific studies of these diseased rabbits were conducted during the 1930s, virus-stricken horned rabbits and hares have long existed. Indeed, representations of the horned rabbit appear in art dating back to the thirteenth century, and during the medieval and early-Renaissance periods the horned rabbit was incorrectly considered by naturalists to be a distinct species.

While rabbits stricken with papillomavirus have been observed in many parts of the US, historically they have been most numerous in the plains states neighboring Wyoming. Douglas—the official "Home of the Jackalope"—is in eastern Wyoming, which puts it close to the point where South Dakota, Nebraska, and Wyoming meet. So, if horned rabbits have long been hopping around Wyoming and its neighboring states, is it possible that the Herrick brothers' first hoax mount may have been inspired by their observation of actual horned rabbits? Or might they at least have heard about such diseased rabbits from other hunters? What about the even more intriguing possibility that in fabricating the first jackalope mount, the Herricks unknowingly invented something that already existed? I am fascinated by the possible relationship between the hoax jackalope and the biological horned rabbit not in spite of the uncertainty that surrounds that relationship, but because of it.

There is no hard evidence to suggest that the Herricks' first jackalope—the one that was for decades displayed in Roy Ball's Hotel LaBonte in Douglas—arose from a source other than the brothers' fertile imaginations. But we do know that horned rabbits attracted curiosity well before 1932, the earliest date given as the birth year of the Herricks' hybrid creature. (Although 1932 was the date claimed by Ralph Herrick, the *Wall Street Journal* had the date as 1934, and a Casper, Wyoming paper reported it as 1939.) According to Peter Jensen Brown's helpful article "Civic Pride through Taxidermy," reports of strange antlered rabbits appeared in newspapers around the country as early as the 1890s, although the term "jackalope" was apparently never used in conjunction with these early notices. Referring specifically to the 1930s, when the Herricks were said to have crafted their first mount, Brown writes that "The earliest-known contemporaneous report of an antlered jackrabbit hunting trophy on display was at a ticket office of the Northern Pacific Railroad in St. Paul, Minnesota, in 1932. And reports made decades after the fact suggest they may have existed during the 1920s, and perhaps as early as 1912."

In fact, the odd phenomenon of competing claims to hoax mounts of other kinds was not unprecedented. During the 1920s the town of Whitefish, Montana maintained that it was the home of the "fur-bearing fish," a precursor to the now widely recognized folkloric "fur-bearing trout" and a fable that, like the jackalope, was anchored by hoax taxidermy mounts. At the same time, however, the residents of Salida, Colorado, objected that their town was the actual home of the furry fish. Whitefish, in turn, may have been inspired by the Great Northern Railway and Glacier National Park which as early as 1913 had attempted to lure tourists with fantastic tales of the apocryphal "polar trout," which was also a type of fur-bearing fish.

Notices of horned rabbits published during the 1930s and 1940s appeared not only in Minnesota, Kansas, Colorado, and Montana, but as far away as Indiana, Michigan, Kentucky, New Mexico, and Texas. For example, I located a brief article from April 16, 1933, byline Texarkana, Texas, with the headline "Horned Rabbit Exhibited." It reads (in its entirety), "Texas, long noted for its cattle, can now boast of rabbits with horns. A rabbit with two horns protruding from the base of the ears was exhibited here by Mrs. Max E. Bertch. It was killed on a farm near Old Boston, Tex." Along with the possibility of real horned rabbits arose a cottage industry to cash in on what was destined to become the most iconic little hybrid in the folklore of the American West. An indicator of the horned rabbit's quickly spreading fame during the Great Depression is the fact that jackalope postcards, which would become be the primary vector of dissemination of the jackalope image, were likely produced as early as the 1930s. But I also discovered a much earlier newspaper notice, this one with a Dallas byline, from September 13, 1915. "Horned Rabbits in Texas" claims that a horned rabbit was not only killed but also skinned and displayed in the office of the local newspaper in Roscoe, Texas, "in order to satisfy those who questioned it."

This same period—the nineteen-teens—also gave rise to a proto-jackalope nickname that would circulate for the next hundred years, even as the roots of the moniker were lost. For years I had come upon obscure references to the horned rabbit as a "War Bunny," with no provenance of the term provided. Most have assumed that the nickname works, as does the slang term "warrior rabbit," to indicate with comic irony that the wee bunny—as the folklore surrounding it often claims—is terribly fierce. However, there is specific historical context for this odd nickname. After much hunting I discovered that the April 1917 issue of *Forest and Stream* (a magazine devoted to outdoor life) ran a short piece called "Enter the War Bunny." In it, author and taxidermist John O'Sullivan related the narrative of a Nebraska hunter who shot a horned rabbit that he brought in to have mounted. "Many there were who doubted the genuineness of the queer looking horns," O'Sullivan wrote, but, after he skinned the strange animal, "no doubt was left about it. The horns were securely and naturally fastened to the rabbit's skull." After describing the specimen in some detail, he went on to speculate that the animal's horns were used for combat, "which has inclined superstitious people who have seen it to the theory that the appearance of 'armed' rabbits indicates that America is destined shortly to enter the arena of war." By the time the *Forest and Stream* issue containing O'Sullivan's notice arrived in subscribers' mailboxes, the horned rabbit's prophecy had been fulfilled. On April 6, 1917, the US declared war on Germany and thus entered World War I. The term "war bunnies" was itself a riff on "war babies," an earlier nickname for children conceived or born during wartime. In the June 1917 issue of the magazine, three readers—one each from Texas, Kansas, and Oklahoma—wrote in to say that they had read O'Sullivan's piece and wanted to offer credible testimony of their own encounters with horned rabbits. The magazine ran this constellation of testimonials under the headline "Where 'War Bunnies' Thrive: In the Lone Star State, Horned Rabbits are so Common Nobody Takes Their Picture."

I have also documented even earlier appearances of horned rabbits. In a 1901 paper, Erwin Barbour, a naturalist who worked for the Nebraska State Museum, described the abnormal growths on cottontails as caused by disease, even as he knew nothing of its pathology. The anomalous bunnies were also mentioned by E. W. Nelson in *Rabbits of North America*, a foundational work published by the US Biological Survey in 1909. That same year, prominent literary naturalist Ernest Thompson Seton included in his book *Life-Histories of Northern Animals* not only a paragraph on horned rabbits but also a sketch of a horned rabbit he had examined. Indeed, Seton apparently did not even consider this anomaly rare, asserting that "Rabbits with horns are frequently found in the dry region of the West." Seton returned to this topic in his 1919 book *Lives of Game Animals*, which includes his hand-drawn illustration featuring four different specimens of "Horned Cottontails."

Although some of these early notices may have been hoaxes, there is ample evidence that hunters had spotted and killed these odd rabbits and sometimes shared their discoveries, at least locally. This is almost certainly why stories of actual horned rabbits were shared through the folk process, making it likely that genuine horned rabbit taxidermy mounts (unlike the hoax mount fabricated by the Herrick brothers) existed before 1932.

So, did the Herrick brothers actually make the first horned rabbit mount? Or did someone else fabricate that first mount—perhaps the one displayed in the St. Paul train station in 1932—and the Herrick brothers heard about it and followed suit? Or did various good-humored taxidermists, completely independent of one another, generate the idea on their own? And there's more to the mystery. If the Herricks did not make the first mount, is it instead possible that the first mount—wherever it might have originated—was no hoax at all, but rather a mount of an actual horned rabbit whose "horns" were caused by viral infection? To thicken the plot even further, might the earliest hoax mounts have been directly inspired by field observations of virus-stricken animals, even if those animals were not themselves made the subject of taxidermy?

The existence of real-life horned rabbits inevitably casts doubt over any horned rabbit hoax mount origin story, leaving us with a delightful mystery—one that seems almost necessary, given that imaginary and fantastic creatures usually emerge from oral narratives, tall tales, and folk art rather than being either proprietary or easily documented. Absent conclusive evidence, pride of place must go to the Herricks and their hometown of Douglas, Wyoming, which not only embraced and actively promoted the horned rabbit, but was almost certainly the source of the moniker that ultimately stuck: *jackalope*. (Compare the Douglas success to the failed attempt by Van Horn, Texas, to promote the town using the "antelabbit" during the 1940s and 1950s.)

President Harry Truman plays a minor role in this strange story. In the spring of 1950 Truman made a cross-country trip by train, ostensibly to celebrate the dedication of several western dams. While the tour was described as "nonpolitical," Truman had a reelection campaign coming up that fall, in which he would defeat Thomas Dewey. Transcripts of comments he made in Casper, Wendover, Cheyenne, Laramie, and Rawlings—some delivered from the rear platform of the train—make clear that he was trolling for Wyoming votes at every whistle-stop. In anticipation of the president's trip, the Douglas Chamber of Commerce issued a series of tongue-in-cheek press releases attempting to draw attention to their town by asserting the local existence of the horned rabbit. One such release, which carried a Douglas byline but was picked up by the United Press, ended up in papers as far away as Paterson, New Jersey, under the headline "Jackalopes Exist, Wyoming Claims."

I have suggested that any attempt to corroborate the facts surrounding the birth of the jackalope is complicated by the existence of actual horned rabbits, which may or may not have inspired hoax mounts. The larger challenge of distinguishing between real and fantastic animals falls not only to biologists, as we might expect, but also to a fascinating group of monster hunters called cryptozoologists. The subject of widespread fascination in popular culture, cryptozoology is the search for and study of animals whose existence remains unsubstantiated. For example, a cryptozoologist might search for an actual species that may or may not have become extinct, like the Ivory-Billed Woodpecker, or one that has been sighted but never captured, like Beebe's Manta, a large, scientifically undescribed ray that has been spotted, photographed, and even videoed in the South Pacific. Or the cryptozoologist might seek an animal not known from the fossil record or by its relationship to existing species, like the Yeti, or other legendary relict hominids. The cryptozoologist often attempts to discover what are called "ethnoknowns," animals unknown to science that nevertheless have a central role in folk legends and myths, particularly within Indigenous cultures (and this because many animals now described by science but once unknown had long appeared in the stories told by native peoples).

In his ambitious compendium *Mysterious Creatures: A Guide to Cryptozoology*, George M. Eberhart offers ten categories into which "mystery animals" can be sorted. His helpful list has been adopted by others seeking an organizational structure by which to classify cryptids. While we tend to think of cryptids in terms of the far-fetched, speculative, and iconic—Bigfoot, Loch Ness Monster, Chupacabra—Eberhart's list broadens our sense of the concept. Among his ten cryptid criteria, or types, are three that apply to the horned rabbit. "2. *Undescribed, unusual, or outsize variants of known species*, such as the BLUE TIGER, HORNED HARE, OR GIANT ANACONDA." The horned hare, one burdened with the growths caused by virus, is undeniably an unusual variant of a known species. "8. *Mythical animals with a zoological basis*, such as the GOLDEN RAM." While we might think of an example like the legendary gryphon, which is derived in part from dinosaur fossils discovered in Central Asia, by this standard the jackalope, too, is a cryptid, because it is a mythical animal that is likely based on the actual horned rabbit found in nature. "10. *Known hoaxes or probable misidentifications* that sometimes crop up in the literature, such as the COLEMAN FROG and BOTHRODON PRIDII." We can agree that the jackalope is also a known hoax. More interesting, however, is the fact that observers unfamiliar with the hoax have sometimes caught a glimpse of an actual horned rabbit in the wild and reported that they had spotted a jackalope. These observers' misidentification occurred because they did not know that horned rabbits exist, or that jackalopes do not.

I do not claim that the jackalope is a cryptid in the same sense as Bigfoot or the Loch Ness Monster—never mind more outrageous creatures such as the Jersey Devil, Mothman, or countless Nessie spinoffs (Chessie in the Chesapeake Bay, Tessie in Lake Tahoe, etc.). But there is value in situating the jackalope as a cryptid. First, because it reminds us that, like many other mythological creatures, the jackalope is likely based on an actual natural phenomenon. Second, because cryptozoology keeps the fantastic and the actual in close conversation with each other, which is a posture well-suited to jackalope hunters. Third, because there is an increasing emphasis on the value of cryptozoology as a way to understand human culture, mythmaking, and psychology—to ask why our monsters are "necessary."

In his insightful article "Cryptozoology in the Medieval and Modern Worlds," Peter Dendle argues that cryptozoology teaches us more about human psychology than about long-sought fabulous monsters:

> The psychological significance of cryptozoology in the modern world has new facets, however: it now serves to channel guilt over the decimation of species and destruction of the natural habitat; to recapture a sense of mysticism and danger in a world now perceived as fully charted and over-explored; and to articulate resentment of and defiance against a scientific community perceived as monopolising the poll of culturally acceptable beliefs. . . . no age has been without its share of hidden creatures, and confirmation of purported species has been a vital and consciously debated issue among the collectors of human knowledge for thousands of years.

As Dendle points out, the imagination of the fabulous animal and the desire to pursue it in order to verify its existence exists across time periods and across cultures. And for good reason, for if the world is either so ecologically impoverished or so thoroughly described that there remains no room for the mysterious and unknown, then we face not only the extinction of species, but also the extinction of an important path of the imagination—one that helps to distinguish us as human. As Peter Steinhart wrote in *Audubon* magazine, "The search for hidden animals is a skirmish in our continuing war against the death of wonder." A related point is made by my friend Robert Michael Pyle, a Yale-trained ecologist, world-class lepidopterist, and prolific environmental writer. In his book *Where Bigfoot Walks* (which was adapted into the feature film *The Dark Divide* in 2020), Bob claims that the protection of wilderness habitats is important for the preservation of our own fantasies and imaginings as well as for the conservation of species. "If we manage to hang on to a sizable hunk of Bigfoot habitat," he writes, knowing full well how unlikely the relict hominid's existence is, "we will at least have a fragment of the greatest green treasure the temperate world has ever known. If we do not, Bigfoot, real or imagined, will vanish; and with its shadow will flee the others who dwell in that world."

George Eberhart's *Mysterious Creatures* begins with an introduction, written by Loren Coleman, entitled "If We Don't Search, We Shall Never Discover." It is a proposition difficult to dispute. Coleman, who is founder and curator of the International Cryptozoology Museum in Portland, Maine (established in 2003), is among the most prominent cryptozoologists in the world. Trained in zoology, anthropology, sociology, and psychiatric social work at Southern Illinois University, Brandeis University, and University of New Hampshire, he is the author of more than forty books on subjects ranging from Bigfoot, lake monsters, and sea serpents to pseudoscience and the subculture of cryptozoology to social welfare issues such as suicide clusters and the so-called copycat effect, a social phenomenon by which violent events widely covered in the media can lead to similar acts of violence.

I tracked Coleman down by phone at the International Cryptozoology Museum—a very real museum that nevertheless has in common with the Museum of Hoaxes and the Merrylin Cryptid Museum that it exists to foster exploration of the liminal zone between the real and imaginary. I had heard that Coleman had a jackalope on exhibit and I wanted to learn what relationship, if any, he saw between the little jackalope and the museum's cryptid superstars, like Bigfoot. When Coleman answered the phone, I was struck by an interesting contrast in the sound of his voice: it had the timbre of an older man's speech, but effervesced with the enthusiasm and energy of a kid's. Coleman explained that he displays his jackalope in a section of the museum specifically dedicated to hoaxes.

"We include hoaxes and fakes to instill critical thinking," he explained. "If someone wants to be involved in cryptozoology, we really want them to think through the question 'Is this creature something that could be a hoax?'" The museum's cluster of deceptions—which is made up of famous taxidermy fakes including the Feejee Mermaid and fur-bearing trout—is specifically intended to make the point that "that's not what we're talking about when we talk about cryptozoology." Coleman chuckled as he explained that his intentions in this respect weren't always successful. In an earlier, smaller iteration of the museum he did not have the fakes clustered together. The result? "I would read visitors' online comments . . . and they would say things like 'I didn't know that the jackalope really existed.' It frightened me. They were missing the whole point! I reorganized it to have all the hoaxes together in one area, with lots of signage."

A veteran monster hunter, Coleman entered the field in 1960, a year before the word "cryptozoology" appeared in print in English. "I've been in the field for sixty years and I've done lots of interviews with the media," he told me. "I often start my interviews saying we in the cryptozoology world do not deal with myths, we deal with legends. Myths come from the human imagination, whereas legends usually have some kind of fire underneath the smoke." Although Coleman viewed the jackalope as a clear fake, rather than as a legitimate cryptid, he did observe some important commonalities between the two. His museum, he explained, was established in response to "two things that people like: animals and mysteries." Jackalopes, like genuine cryptids, check both of those boxes, and he has seen the powerful effect they can have on visitors. "Even though people can hear the explanation—this was made, we got this at Cabela's, or something like that—they're still saying there's a mystery here, maybe they really are out there. They move the jackalope into the cryptid world even though they really are not, so the jackalope applies to cryptozoology for some people touring the museum." Then he added a more global comment that I found fascinating: "I'm very secure in understanding that cryptozoology is part of folklore."

I asked Coleman how visitors respond to the museum's jackalope when they see it. "People tend to have very joyful reactions," he replied, obviously joyful himself. "They're very happy. They sort of feel like they've been looking for jackalopes all their life, and now they've found one. Which is phenomenal!"

"Well," I observed, concluding our very interesting conversation, "it sounds like we agree that, however the jackalope is classified in relation to cryptids, it has captured our imagination." I find myself thinking again of Jorge Luis Borges's concept of the "necessary monster."

"Oh, yes, definitely," replied the famous cryptozoologist. "I see it as a real fulfilment of a human need. Humans *need* jackalopes."

Chapter 7
The Global Jackalope

Why pick up rabbit horns in the desert,
Or knit a sweater out of turtle hair?
You have to be blind to see this path
And deaf to hear these instructions.
Sketch a scene on the side of a wave.
Climb a staircase in a dream.

—Ken McLeod, *An Arrow to the Heart: A Commentary on the Heart Sutra*

When I was a kid, back in the dark ages before the dawn of on-demand programming, Saturday mornings were devoted to watching cartoons. Of the nearly four hundred characters voiced by the great Mel Blanc—among them Daffy Duck, Porky Pig, Yosemite Sam, Foghorn Leghorn, and Wile E. Coyote—the greatest of all was Bugs Bunny. Although the Bugs Bunny cartoons ran from the wily rabbit's debut in 1940 through his swan song in 1964, in syndication they remained standard fare for decades. The iconic star of the Warner Bros. *Looney Tunes* and *Merrie Melodies* series, Bugs was a one-of-a-kind character whose exploits not only entertained, but also provided a vicarious thrill for any kid who felt insecure, bullied, or simply constrained by the oppressive demands of good behavior imposed on them by the adult world. To understand Bugs's charm was to comprehend the appeal of a trickster rabbit who is a cousin of the jackalope.

Bugs was as smooth as Clark Gable, whose nonchalant, fast-talking, carrot-munching mannerisms in *It Happened One Night* (1934) provided inspiration for his character. His one-liners were as droll as those cranked out by Groucho Marx, another inspiration for Bugs—though Bugs gnawed his carrot in place of Marx's trademark stogie. In addition to being cool and comical, Bugs was impressively self-reliant and profoundly anti-authoritarian, standing up for himself no matter who tried to shoot, eat, evict, capture, or co-opt him. Constantly hunted by Elmer Fudd and menaced by Yosemite Sam—two heavily armed humans who nevertheless remained unable to best him—our underdog hero prevailed against all odds. If he wasn't physically stronger than jocks (*Baseball Bugs*, 1946), bullies (*Rabbit Punch*, 1948), gorillas (*Gorilla My Dreams*, 1948), pirates (*Buccaneer Bunny*, 1948), bulls (*Bully for Bugs*, 1953), robbers (*Bugsy and Mugsy*, 1957), yetis (*The Abominable Snow Rabbit*, 1961), or bears (*The Iceman Ducketh*, 1964), his quick wits proved more powerful than the brute strength of any adversary.

Delightful animation, excellent storytelling, superbly voiced characters, eye-popping technicolor: all of these factors contributed to the allure of the Bugs Bunny cartoons. Far more important to me as a kid, though, was the fact that a bunny—an animal associated with timidity and fearfulness—was an absolute ass kicker. The drubbings he delivered were made possible by characteristics I feared lacking in myself: confidence, creativity, resourcefulness, ingenuity, and verbal aptitude. Bugs was a cunning shape shifter who donned wild costumes, adopted foreign accents, played myriad roles, and was always a smart-ass of the first order.

Like most trickster figures, Bugs Bunny was not without character flaws. For example, he often deployed dishonesty and deception—the very behaviors we kids were warned against. But this is precisely why the wascally wabbit seemed to me a rebel hero. While I was usually found out and punished for my own rascally deeds, Bugs not only got away with socially unacceptable behaviors, he prospered as a result of them. While I rationalized that the crafty rabbit's atrocious conduct was morally justified because it was often employed in self-defense, the truth is that I simply found his contemptible antics irresistible. Meanwhile, the connection between the little rabbit and the skinny kid sitting cross-legged too close to the TV could hardly have been more obvious: Bugs's resilience was my strength, his triumph my victory, his endurance an assurance that I would survive to adulthood.

In addition to being anti-authoritarian, slick, and hilarious, Bugs Bunny was, like other folk heroes, larger than life, magically unbound by the limitations that constrain the rest of us. In *High Diving Hare* (1949), for example, he breaks the fourth wall (as he so often did), addressing the viewer directly to explain why even the laws of physics do not apply to him: "I know this defies the law of gravity, but I never studied law." My favorite example of Bugs's tall-tale strength is *Rebel Rabbit* (1949), an episode whose plot is initiated when an affronted Bugs discovers that the two-cent bounty on rabbits is so pitiably low because rabbits pose little threat to property and are thus considered less "obnoxious" than animals like foxes and bears. Setting out to prove that "a rabbit can be more obnoxious than anybody!" Bugs paints barber pole stripes on the Washington Monument, hacks the lights in Times Square to read "BUGS BUNNY WUZ HERE," shuts off the flow of Niagara Falls, saws Florida off from the rest of the country, and, grabbing a shovel, fills in the Grand Canyon. ("Well," he remarks casually, "that fills up that hole.") Was Bugs obnoxious? Absolutely! But he was as powerful as any Paul Bunyan, Pecos Bill, Davy Crockett, or John Henry could ever be, while also being a lot funnier. Bugs's epic humor even had philosophical depth, as in his existentialist quip, "Don't take life too seriously. You'll never get out alive" (*Rabbit's Feat*, 1960). An iconic trickster, Bugs functioned as an animated version of a jackalope: something that wasn't real, but could fool people just the same.

In January 1961, Mel Blanc—the mortal voice of the immortal Bugs Bunny—was in a near-fatal head-on car crash that left him in a coma for two weeks. The story of how the celebrated voice actor reemerged from oblivion is, like the tale of the jackalope, the stuff of legend. The doctor, trying to prompt his unresponsive patient, asked, "Bugs Bunny, how are you doing today?" Blanc suddenly answered, in Bugs's voice, with his trademark line: "What's up, Doc?"

Tricksters, including Bugs Bunny, have been with us from time immemorial. They are shape-shifters, boundary crossers, resilient, resourceful, clever, even magical figures whose unconventional behavior defies social norms. The trickster's deviant behavior releases a repressed energy the community needs, and so the trickster often functions as an intermediary between our mundane world and difficult-to-conceive realms beyond our own. This intercessory role is one reason tricksters are often conceived as animals—those coyotes, ravens, and spiders who regulate imaginative commerce between human and nonhuman worlds. Terri Windling, in her discussion of the cross-cultural iconography of rabbits and hares, writes that "In some lands, Hare is the messenger of the Great Goddess, moving by moonlight between the human world and the realm of the gods; in other lands he is a god himself, wily deceiver and sacred world creator rolled into one."

To understand the mythical power of the horned rabbit, it is helpful to acknowledge how widely the figure of the trickster rabbit appears across world cultures. Like our pal Bugs, rabbits and hares are often shape-shifters and tricksters who hold a special place in mythologies from around the globe. Consider, for example, the association of rabbits with the lunar cycle. In China, the Hare in the Moon (what western folklore construes as the Man in the Moon) is represented by a mortar and pestle in which he blends a potion that ensures immortality. Variants of the Moon Rabbit also occur in Japanese and Korean mythology. In ancient Egypt, temple walls were adorned with dramatic images of the hare-headed goddess Wenet and her male companion Wenenu, who may have been thought of as a form of Osiris, the god of death and resurrection. (The Egyptian word for hare, *un*, meant "the opener" or "to open," the etymology of which may be linked to the fact that hares are born with their eyes open.) And in Ugrian mythology, Kaltes-Ekwa, goddess of fate, often appeared in the form of a hare.

The more global trickster hares I encounter, the stranger and more beautiful their stories become. The Centzon Tōtōchtin of Aztec mythology was a group of divine rabbits who gathered frequently to throw drunken parties. In Skandanavian folklore, the Milk Rabbit sucked milk from the udders of cows and spit it into pails carried by witches. *Kojiki*, an eighth-century Japanese chronicle, tells the marvelous tale of the sacred Hare of Inaba. The hare, needing a means to cross from the Island of Oki to Cape Keta, challenged the sharks to see whether the clan of the sharks or the clan of the hares was more numerous. The trickster hare instructed the sharks to line up in a row across the sea, then he hopped across them, purportedly to count them, but actually to deceive them into serving unwittingly as a land bridge for his crossing.

I also find it fascinating that trickster rabbits are often represented as divine. Ostara, the goddess of the moon in Anglo-Saxon myth, was portrayed with rabbit's ears. Ostara's Celtic incarnation, Ēostre, magically transformed her pet bird into a rabbit that laid brightly colored eggs which she gave to children, and thus may be an ancient link to some modern traditions associated with Easter. Among Celtic tribes the eating of rabbits was forbidden, probably because shape-shifting goddesses so often transformed themselves into rabbits. The Celts even used rabbits for divination, studying the patterns of their tracks and searching for signs within their entrails. They also believed that the rabbit's burrow was a passageway to the spirit world, and that rabbits were empowered to shuttle messages between the living and the dead.

These are just a few examples of the vital folkloric and spiritual significance rabbits and hares have held in cultures around the globe. What is more surprising is how many cultures specifically feature *horned* rabbits in their mythology and folk narratives.

Among the most ancient of these mythical horned rabbits is the Al-Mi'raj (in Arabic, أَلْمِعْرَاج), described by thirteenth-century Persian naturalist Abu Yahya Zakariya' ibn Muhammad al-Qazwini in his geographical encyclopedia, *'Ajā'ib al-makhlūqāt wa gharā'ib al-mawjūdāt* (*Marvels of Things Created and Miraculous Aspects of Things Existing*). Among the most influential cosmographies to emerge from Islamic culture, al-Qazwini's book survives in many copies and in numerous translations from Arabic into other Islamic languages. In the book's chapter on the China Sea, al-Qazwini describes the Al-Mi'raj as "a yellow beast that looks like a rabbit and has a black horn." According to al-Qazwini, an island called Tinnin was terrorized by a huge dragon that ate the people's cattle and, if they approached too closely, devoured the people themselves. In order to appease the monster, the island's inhabitants offered two oxen each day as tribute. When Alexander the Great's epic voyages brought him to Tinnin, the islanders begged him for help. The Greek king first ordered that two oxen be killed and skinned. A trickster himself, Alexander stuffed the skins with arsenic and other poisons and, for good measure, iron hooks. (His ingenuity may be less surprising when we remember that he was tutored by Aristotle.) The unsuspecting dragon consumed the bait, retreated to its lair, and was soon after found dead. In appreciation for their deliverance from the beast, the people of Tinnin made Alexander a gift of the strange rabbit-like animal with the single horn. Captivating depictions of the creature may be seen in various illustrated folios of al-Qazwini's book.

Deep in the cloud-shrouded forests of central Mexico's Sierra Madre Occidental Mountains live the Huichol, an Indigenous people who have occupied the region for at least 15,000 years. Huichol oral narratives often feature the deer god Kauyumári, "a divine culture hero, guide, and messenger between the worlds of mortals and the deities." Over a century ago, the German ethnographer Konrad Theodor Preuss collected a Huichol myth in which, during the legendary "First Times," both the deer and rabbit had horns—though the deer, seeking to be the only horned animal, ultimately tricked the rabbit into giving his up.

Huichol storyteller Ramón Medina Silva shared several variants of this narrative with other folklorists in 1966. His stories, Silva says, come from "a time when rabbit was a deer." In the first of these, Deer loans his antlers to Rabbit. However, while swimming in the river, Rabbit finds the loan a burden: "'They are too heavy for me,' Rabbit says, 'my head goes under.' So Elder Brother [Deer] relieves [Rabbit] of the antlers and puts them back on his own head. 'They properly belong to you,' Rabbit says. 'They do not look good on me.'" In Silva's second version of the story, Rabbit instead wins his antlers in a wager. Deer challenges Rabbit to best him in leaping over the water, and offers to give Rabbit his *muwieri* (ceremonial arrows that symbolize antlers) should he win. Rabbit is victorious, and receives his rack. However, the prize proves too heavy for Rabbit, who ultimately returns the antlers to Deer, where they have remained ever since. As Pre-Columbian culture specialist Jill Leslie Furst explained, "The Huichols insist that formerly rabbits could be misidentified as deer because of their shared antlers." Similar horned rabbit ancestors exist in the oral lore of other Indigenous Mesoamerican cultures, as the deer and rabbit have often been closely associated—sometimes as a pair of brothers, or as the sun and moon—as far back as Aztec culture, which flourished at about the same time Abu Yahya Zakariya' ibn Muhammad al-Qazwini's tales of the Al-Mi'raj were emerging from the Islamic world.

If the horned hare is an important character in native Huichol mythology, he is the star of the show in the folklore of many African cultures, including the Ila-speaking peoples of Zimbabwe, the Ronga of Mozambique, and the Mandinka of the Gambia. Many of these African stories concern a trickster rabbit who cunningly fabricates horns and dons them as a ploy. In one interesting variant, told in Kudang Village in the Central River Region of the Gambia, the hyena informs the hare of a party to occur that night, adding that, much to his disappointment, only animals with horns are invited. Promising to help the hyena, the hare gathers some branches from the forest and sharpens them with his teeth. But when the hyena bows his head to receive the false horns, the hare positions the sharp branch behind the hyena's ear and drives it into his skull with a rock. Having killed the hyena, the hare brags of his deed to the antelope, who steps out from the party to discover that the hyena is no more. "'You must be very clever indeed,' said the Antelope, 'to have bested one so ferocious as the Hyena. Please come in and tell us all about it.'" Thus it is the trickster hare, rather than the hyena, who is ultimately allowed to attend the exclusive gathering.

A more common version of the African horned hare story is related by storyteller Samuel Mabika, of the Mozambican Ronga people. In Mabika's tale, Ňwampfundla, the hare, is servant to the lion, who is the great chief. Traveling with an entourage of his animal servants, one evening the lion decrees that the fruit of a certain tree is to be reserved for him alone. Trickster that he is, Ňwampfundla hatches a plan for how he might eat the fruit and avoid being caught. While the other animals slumber, the crafty hare consumes the fruit from the forbidden tree, then gathers the fruit pits into a bag, which he hides with the sleeping elephant in an attempt to frame the blameless pachyderm for his own misconduct. When morning comes and the bag of pits is discovered, the innocent elephant pays for Ňwampfundla's crime with his own life. After the hare's wrongdoing is discovered, he takes refuge in his burrow, from which he taunts the lion and the other animals who seek justice for the elephant's death. When the lion and his crew finally leave, the hare sneaks out to play yet another trick. "Then he went to a beehive that he found in a tree nearby and took some of the wax that was there and made two little horns of that wax and put them on his head, so that it appeared that he had two little horns growing on his head." The false horns help Ňwampfundla to disguise himself, after which he has the audacity to return to the lion, who fails to recognize him. The lion, who is "pleased to see this new kind of hare" (a horned hare), unknowingly welcomes Ňwampfundla back into his entourage. As is common with trickster figures, the hare benefits from his dissembling and is never brought to justice—though in a Zimbabwean variant of this tale the hare approaches too close to the fire, his wax horns melt, and his ruse is revealed.

One of the many blessings of being widely known as "the jackalope guy" is that I receive fascinating and often entertaining horned rabbit-related correspondence from friends and strangers around the world. In early 2019, I received the following message from my friend Lauren LaFauci, an American who lives in Sweden, where she teaches environmental humanities at Linköping University:

> In late 2018, my colleague Cecilia Åsberg and I met at her home to talk through our project collaboration plans for the coming year. I was standing in her doorway when I noticed the jackalope. Since I married into a Wyoming family, I was very familiar with this creature and its status in the Mountain West, but here I was on the flatlands of Östergötland in eastern Sweden, some 5,000 geographic (and perhaps ten times as many cultural) miles away.
>
> "Cissi! Is that a jackalope?" I was shocked. On her wall was the taxidermied head of a jackalope, colorfully mounted on a wooden, turquoise plaque. She replied that no, it was a "råkanin," of course.
>
> The second part of that word was easy for me, now three years into my study of the Swedish language: "kanin" simply means "rabbit" in Swedish. It was the first part that piqued my interest, though. "Rå" can mean a variety of things, including "raw," but a "raw rabbit" made little sense in this context.
>
> "Rå," when linked to the word for animal ("djur"), also means "deer," so a "deer-rabbit" seemed more on track: here was a rabbit's face with a deer's antlers. But take it one layer deeper and the Swedish word "rå" gives us a more evocative reading: in Swedish folklore, "rå" refers to a spirit inhabiting a particular place. You can have "skogsrå" (forest spirits), "sjörå" (lake or sea spirits), "bergsrå" (mountain spirits), and so on. I would guess that the word for "deer" likely stems from this meaning of "rå": in inhabiting the forest so lightly, so gracefully, deer

are, in Swedish, literally, "the spirit(s) of the forest." Then, too, we can see how "raw" could also mean primal, natural, close to origins. The Swedish "råkanin" thus holds evocative layers both etymologically and folklorically.

Cissi had never heard the word "jackalope" prior to our meeting, nor I the word "råkanin." But together we wondered, had Swedish immigrants to the American West taken this creature's stories with them? I recalled the ruins of Swedish-built log cabins I had seen on hikes high in the Uinta Mountains of northeastern Utah, near the Wyoming border, in land now designated "wilderness." Surely, forest spirits or mountain spirits live there, too. Or, had American culture infused the Swedish in the other direction: did a Swedish traveler bring jackalope stories or objects back from the US as an appealing piece of twentieth-century American kitsch?

Lauren, Cissi, and I were never able to unlock the mystery of the råkanin. Had it migrated east across the Atlantic, perhaps from the wilds of the American West? Was it a forest spirit, part of a horned rabbit tradition that developed independently in Sweden? Or, perhaps, was this little råkanin inspired by a virus-stricken rabbit that had appeared in the Swedish flatlands way back when? It is a predicament I have found myself in many times. Even for a tireless jackalope hunter like me, tracking hybrid and horned rabbits across world cultures is like playing a game of folkloric hide and seek. Mythical rabbits, rabbit deities, antlered rabbits, every kind of magical hare imaginable—these cross-cultural jackalope cousins pop up, demand my attention, and then vanish into their warrens before I can grab them by the ears (or horns). Information about these folkloric bunnies is often scarce, sometimes unreliable, and usually confined to native languages. (My jackalope expedition has required that I attempt, with a great deal of help, to translate Spanish, French, Italian, German, Dutch, Swedish, and Arabic). Any quarry can be elusive, but when you are pursuing animals that are mythical, imaginary, and sometimes played as hoaxes, it is common that the evidence turned up is fragmentary or dubious.

To give a sense of how many horned rabbit trails cross in the dark forest, consider the sole case of Germany. For starters, there is the dilldapp of the Alemannic region, which is reported to inhabit the Hauberg (communally maintained forest) near the town of Siegerland, in North Rhine-Westphalia. A bizarre hybrid of a skunk, rabbit, and deer, this strange animal is so revered in the region that members of some fools' guilds dress up as dilldappen as part of their reveling and festivals. Legend has it that the dilldapp is quite shy, and feeds mainly on potatoes. Handy, however, is the fact that the animal's tongue is so raspy that it can be used to grate those potatoes. When confronted, the dilldapp performs the "Siegerland tongue strike," by which it shoots its tongue out of its mouth at the attacker. The male dilldapp, called a "glonk," weighs up to 4 kilograms. The unhappily named female, the "dildo," lays eggs only once or twice in a lifetime. This, however, is sufficient to sustain the population, since dilldappen live a century or more.

Then there is the hasenbock (or hirschhase), a "billy hare," or "deer hare," which is essentially a lesser-known species of a larger genus of German *gehörnte hasen* (horned hares). The rasselbock (or raspelbock), a more prominent relative associated with the Thuringian Forest in southern Germany, is an animal with the head and body of a rabbit and the antlers of a roe deer. As with the elusive snipe and jackalope, the ritual hunting of the rasselbock is a key part of the folklore surrounding it. Traditional practice at hunting lodges in the region is to persuade the uninitiated not only that the rasselbock exists, but that it can be hunted. The group leaves the lodge for the forest, where they station the dupe with a sack, in which he is to capture the animal as it is driven toward him. The seasoned hunters then disperse into the forest, presumably to scare the animal toward the uninitiated hunter. Instead, the sucker is left literally holding the bag, as the experienced hunters retreat to the lodge to await the return of their victim who, upon realizing the trick, will be welcomed with laughter and a round of drinks. So beloved is the "rasselbockjagd" (rasselbock hunt) that the town of Blankenhain, in Thuringia, has issued mock bank notes featuring the animal as an unofficial town mascot. The front of the bill depicts a cavorting pair of horned, rabbit-like beasts. The reverse side of the currency depicts the moonlight scene of the hunt, as several men wearing overcoats and Tyrolean hats carry a lantern, club, and burlap sack. The German text accompanying the image reads: "Germany's hunting grounds are large, but rasselbocks are only found here—One catches them by the light of the full moon, alive, inside of a sack." Each spring, after the inaugural rasselbockjagd, the town celebrates the Rasselbock Festival, complete with a feast and the ceremonial crowning of Miss Rasselbock.

The star of German *gehörnte hasen*, however, is the wolpertinger (or wolperdinger, woiperdinger, volpertinger, or woibbadinga), a beast that has the head of a rabbit and the antlers of a deer, but also sometimes the wings of a pheasant, and/or the fangs of a dwarf boar. Here, too, regional variation occurs. In Oberpfalz the wolpertinger is called the oibadrischl; in Niederbayern it is known as the rammeschucksn; over the border in neighboring Austria it is referred to as the raurakl. According to the folklore of southern Germany, the wolpertinger roams the alpine forests of Bavaria and Baden-Württemberg. Reputed to be a shy but mischievous creature, the wolpertinger has long been a staple of Bavarian mythology. Seventeenth-century engravings and woodcuts depict the hybrid animal, which was "made real" by enterprising Bavarian taxidermists during the eighteenth century. Throughout the nineteenth and twentieth centuries, the animal appeared in restaurants, hotels, and hunting lodges, where its display enchanted locals and fooled outsiders. In all these respects the wolpertinger is a close analogue of the jackalope. Also like the jackalope, a rich body of lore accompanies the wolpertinger. For example, it is said that if its saliva touches human skin, bushy tufts of hair will sprout. If the animal feels threatened, it will spray a foul-smelling liquid, leaving a stench that cannot be removed with any soap or deodorant; however, the stench will magically disappear exactly seven years after the encounter.

Hoping to learn more about this remarkable creature, I struck up a fruitful correspondence with Dr. Anne Blaich, of the *Deutsches Jagd-und Fischereimuseum* (The German Hunting and Fishing Museum), in Munich, Germany. Anne confirmed that wolpertingers have been exibited at the museum since the 1950s. When I asked her if the hoax mounts actually fooled visitors, she replied (in English far better than my German) that it depended on who was doing the visiting:

> It is noticeable among the foreign visitors that especially the Russian visitors like the Wolpertinger and regard it as a real animal. Asian guests are also very impressed by the Wolpertinger, but they know that it is a mythical creature. And with the German visitors you have to differentiate between locals and foreigners. Those who are not Bavarians do not believe that Wolpertingers are real animals. In Bavaria, Wolpertingers are part of our culture. It is a being that the Bavarians are proud of and can identify themselves with. Children in Bavaria are told about Wolpertinger like they get told about the Easter Bunny and Santa Claus. Wolpertinger are real in the Bavarian fantasy, but older children and adults know that the Wolpertinger is not a real animal.

In rereading Anne's message, she seemed to be saying that only in Bavaria did people believe in the existence of an animal they knew to be imaginary—a paradoxical insight very much at the heart of how those of us living in the American West feel about our jackalope. Only an outsider would be so foolish as to believe that the jackalope exists, or so jaded as to maintain that it does not.

The importance of the mythical wolpertinger to regional cultural identity became increasingly clear as I continued to exchange messages with Anne. She explained that in the past twenty years in Upper Bavaria, there has been a concerted effort to celebrate old customs and myths. Citizens of the region have begun to wear traditional dress and to preserve the

Bavarian language. As part of this effort to revive and transmit local customs, Bavarian kids are now learning about the wolpertinger in school.

I asked Anne if there was such a thing as a wolpertinger hunt, as there is with the jackalope and rasselbock:

> Oh, yeah, there is! There are different versions but they all have the following in common. If you want to catch a Wolpertinger, the hunter and a pretty young woman have to go hunting under the full moon. You will need a sack, a candle, and a stick. The sack is placed on the ground. The stick is placed in such a way that it holds the sack open. Then the candle is lit and placed in the opening. The Wolpertinger is attracted by the light and the scent of the beautiful woman. He crawls into the sack. Now the hunter can close the sack quickly and the Wolpertinger is caught.

The hunt is played ritually as a gag on the unsuspecting visitor or novice hunter and is a reliable source of good fun among Bavarians.

In closing my correspondence with Anne, I asked whether visitors who experience the wolpertinger are familiar with its American cousin, the jackalope. "We Bavarians consider ourselves a special people," she replied. "Lovable, solitary, even self-ironic, but also somewhat arrogant toward strangers. The Bavarians are therefore convinced that such a special animal as the wolpertinger can live only in Bavaria."

Lepus cornutus, the Latin name for "horned hare," has analogues in many European languages: *lièvre cornu* in French; *lepre cornuta* in Italian; *gehoornde haas* in Dutch; and *gehörnter hase* in German. As the binomial nomenclature of the name indicates, *Lepus cornutus* was once believed to be a distinct species, and it was described and depicted by European naturalists during the Renaissance. Starting in the late sixteenth century and continuing through the early nineteenth century, *Lepus cornutus* appeared in the text and illustrations of some of the period's most significant natural histories.

One of the last important manuscript illuminators, the Flemish painter and draftsman Joris Hoefnagel, specialized in natural history subjects and topographical views. He took as his personal motto *natura magistra* ("nature his teacher"). Hoefnagel's highly realistic natural history illustrations were collected into a four-volume manuscript (with folios dated from 1575–1582) now known as the *Four Seasons*. In the volume *Animalia quadrupedia et reptilia (terra)*, Hoefnagel included a remarkable plate depicting a squirrel on a branch extending above three rabbits (folio 47, plate 4). The rabbit on the right appears perfectly normal; that on the left is normal also, save for the conspicuous fact that it is a replication of one of the most famous rabbits in the history of art: German Albrecht Dürer's 1502 watercolor *Young Hare*. The odd rabbit in the starring role at the center of the illustration, however, sports a fine pair of branching antlers.

Many seventeenth-century naturalists followed Hoefnagel's lead. Florentine engraver Antonio Tempesta depicted a pair of hares with gently curving horns in *Nova raccolta de li Animali li piu curiosi del mondo* (circa 1600), while prominent Swiss physician and naturalist Conrad Gessner included a detailed sketch of a horned rabbit's skull and horns in his *Historia Animalium Quadrupedum* (1602). In *Animaliam Quadrupedum Vivae Icones* (1612), Flemish designer and engraver Adriaen Collaert depicted a large horned hare facing a fox in a posture reminiscent of the allegorical style of medieval bestiaries. Ulisse Aldrovandi, the Italian naturalist who helped create Bologna's remarkable botanical gardens (among the first in Europe), depicted a pair of horned hares in *De Quadrupedibus digitatis viviparis libri tres* (1637). These hares, which closely resemble those created by Tempesta, are included on a page with two other species of *Lepus*, making clear that, like most naturalists of his age, Aldrovandi considered the horned hare to be a distinct species. The Polish scholar and physician John Jonston's *Historiae naturalis de quadrupetibus libri* (1655) includes illustrations of horned rabbits similar to those produced by Aldrovandi. German Jesuit and scientist Gaspar Schott borrowed illustrations of *Lepus cornutus* from Jonston and included them in his *Physica Curiosa* (1662), while English parson-naturalist John Ray mentioned the horned hare in *Observations topographical, moral, & physiological made in a journey through part of the low-countries, Germany, Italy, and France* (1673). In a 1687 article, which included an illustration, German alchemist Gabriel Clauder claimed to have seen a horned hare himself.

At the same time these seventeenth-century naturalists were describing and depicting what they considered a unique species of *Lepus*, the horned hare also proved an enticing subject for artists. For example, early in the century a horned hare was painted by an anonymous artist in one of the coffers of the ceiling of the gallery of Chateau d'Oiron (Deux-Sevres), a castle in western France. Around 1620, influential Flemish painters Jan Brueghel the Elder and Peter Paul Rubens collaborated on a remarkable painting called *The Virgin and Child in a Painting surrounded by Fruit and Flowers* (aka *Madonna and Child in a Garland of Fruits and Flowers*), for which Brueghel painted the settings, garlands, and animals, while Rubens created the human and angelic figures. This tender and beautiful work represents Mary cradling the Christ child as she prepares to be crowned by a pair of hovering cherubim. In the foreground right of the painting sits a handsome horned hare. Another Flemish painter, Jan van Kessel the Elder, also depicted horned hares in *The Four Parts of the World* (circa 1660s).

That horned rabbits had long been the subject of more frivolous and ephemeral forms of visual art is suggested by a revealing passage from the Prologue to Book I of French polyglot François Rabelais's monumental satirical novel *Gargantua and Pantagruel*, which dates to 1532. In it, Rabelais explains to his readers what a "Silenus" was: "those used to be little boxes, the kind you see, today, in drugstores, painted all around with light and happy figures, like harpies, satyrs, bridled geese, hares with horns, saddled ducks, flying goats, stags in harness, and all sorts of such images, invented in good fun, just to make the world laugh." While many representations of *Lepus cornutus* from this period are preserved in archives specializing in early natural science, the horned hare's aesthetic and imaginative qualities led to the figure appearing in art now held at the Louvre in Paris, the Prado in Madrid, the British Museum in London, the National Gallery of Art in Washington, D.C., and other repositories of fine art around the world. Clearly, the jackalope's ancestor had captured the imagination of seventeenth-century European scientists, artists, and writers.

Even as natural science matured during the eighteenth century, *Lepus cornutus* continued to make appearances in natural history texts across Europe. German botanist and zoologist Jacob Theodor Klein included the would-be species in his *Summa Dubiorum Circa Classes Quadrupedum et Amphibiorum* (1743), as did Klein's countryman, the engraver and printmaker Johann Daniel Meyer, whose *Angenehmer und Nuetzlicher Zeitvertreib mit Betrachtung Curioser* (circa 1750) included a pair of highly stylized horned hares with bold stripes and fantastically branching antlers. By the mid-eighteenth century, fewer naturalists were willing to accept *Lepus cornutus* as a distinct species, and most came to view the animal as a freak of nature, though they had no understanding of the pathology causing the growth of the animal's horns. For example, Johann Elias Ridinger, the famous German engraver of hunting scenes, in captioning his depiction of the "Hirschhase" or "Deer Hare" (circa 1760), refers to it not as a unique species but rather as a *Naturseltenheit*, or "natural rarity."

Georges-Louis Leclerc, Comte de Buffon, a monumental a figure in eighteenth-century natural history, was also curious about the horned hare. In the 1770 edition of his *Of the Degeneration of Animals*, Buffon wrote that "It is claimed that there are in Norway and in some other northern provinces hares which have horns," but added that he considered their existence uncertain. Buffon speculated that "this variety, if it exists, is only individual and probably only occurs in places where the hare does not find any grass, and can only feed on woody substances." Similar hedging occurs in von Johann Christoph Heppe's *Systematisches Lehrbuch über die drey Reiche der Natur* (1778). Under the genus *Lepus* he included "Der Gehörnte

Hase" as a species, but added a qualifying note: "The horned hare. This is presumably just a *naturspiel*" (a trick or freak of nature); that is, the horned rabbit is not a species but rather an anomaly.

Even as scientists came to understand that the horned hare was not in fact a distinct species, the freak animal continued to be an object of fascination that was featured in natural history texts of the late eighteenth and early nineteenth centuries, including French engraver Robert Bénard's plates, which appeared in Pierre Joseph Bonnaterre's *Tableau Encyclopédique et Methodique* (1790), German naturalist Johann Christian Daniel von Schreber's *Die Säugthiere in Abbildungen nach der Natur mit Beschreibungen* (circa 1792), and German copper engraver Jacob Xaver Schmuzer's plate in Friedrich Justin Bertuch's *Porte-feuille instructif et amusant pour la jeunesse* (1807).

So, horned rabbits have appeared in sources as varied as a thirteenth-century Persian cosmography, oral traditions of the Indigenous Huichol people of Mexico, folktales from several regions of Africa, creation myths of some native peoples of North America, mythology from various European cultures (including a spate of regional variants in Germany alone), not to mention a satirical novel by Rabelais, a devotional painting by Brueghel and Rubens, and a remarkable number of natural histories by prominent early scientists across Europe. It is clear that the jackalope's global precursors are impressively numerous.

Where, then, are all these horned rabbits coming from? Although variation in how the animal is described and depicted suggest that early horned rabbit stories and images were not always created in the presence of physical specimens, some of the descriptions and images do accurately represent the growths on actual rabbits stricken with papillomavirus. This fact, along with the prevalence and wide dissemination of the horned rabbit across cultures, leads me to believe that in many cases the jackalope's international cousins are indeed *naturseltenheit*: actual natural rarities rather than fantasies or hoaxes. Because horned rabbits exist in nature, observation of them has almost certainly been the reason for their representation in many cultures over many centuries. Is it possible that some of these jackalope analogues originated only in the folk imagination, and were not inspired by observations of actual horned hares? Certainly. But while the wolpertinger is an obvious deception and *cornutus* is not a distinct species of the genus *Lepus*, the fact that the horned rabbit appears so widely in stories, texts, images, and, most importantly, early scientific documents, suggests that its real-world presence is the inspiration for many of these globally disseminated jackalope cousins.

Join me as I conclude my international horned rabbit expedition with the oldest and strangest jackalope precursor of all. This elusive horned hare is one I am still chasing, and will be for a long time to come.

The ancient Jātaka tales, which emerged in India from 300 B.C.E. to 400 C.E., are stories concerning the many previous lives—both human and animal—of the Buddha. In one of these, called "The Selfless Hare," Buddha was reborn as a hare who lived in the forest with his three friends, the monkey, the jackal, and the otter. One holy day, Śakra, Ruler of the Heavens in Buddhist cosmology, looked down from the summit of Mount Meru and decided to test the animals' virtue. Now, the animals had been out foraging: the otter had found several red fish on a riverbank; the jackal had caught a juicy lizard; and the monkey had deftly harvested ripe mangoes from the treetops. Disguising himself as a Brahman (priest), Śakra went to each of the animals in turn, saying, "I am hungry and must eat before I can perform my priestly duties. Can you help me?" The otter immediately shared all his red fish. The jackal gladly gave the Brahman his lizard. The monkey did not hesitate to hand over his mangoes. But, when asked, Buddha (in the form of the hare) replied that he had no food and instead requested that the Brahman build a fire. When the fire was burning hot, the hare exclaimed, "I have nothing to give you to eat but myself," and with that he leapt into the fire. Śakra was so moved by the selfless action that he made the fire go cold in order to prevent the hare from burning to death. "Dear hare," said Śakra, revealing his true identity, "your virtue will be remembered through the ages!" To commemorate the hare's inspiring act of self-sacrifice, Śakra painted the hare's likeness on the face of the moon, where it may be seen today.

Even more fascinating to me than Buddha's reincarnation as a shape-shifting self-sacrificing hare is the fact that Buddha himself used the horned hare as a potent metaphor—and as a teaching tool to help his students think deeply about the nature of reality and the mind. Buddha's horned hare appears in texts far more ancient than that written by thirteenth-century Persian naturalist al-Qazwini. I have discovered the horned rabbit lurking in three Mahayana Buddhist scriptures, or sutras: the Platform Sutra (circa 820 C.E.), the Surangama Sutra (circa 705–730 C.E.), and the Lankavatara Sutra (circa 350–400 C.E.).

The Platform Sutra, which is especially important in the Zen tradition, includes a *gatha* (the versified part of a sutra) that features this fascinating stanza:

> One who wishes to teach others
> Must have skillful means.
> Do not allow them to have doubts.
> Their own intrinsic nature will then appear.
> The Dharma [divine law] is within the world,
> Apart from this there is no awakening.
> Seeking *bodhi* [enlightenment] apart from the world
> Is like looking for a rabbit's horn.

The core message of this intriguing passage seems to be that spiritual illumination should be sought not in transcendence, but rather through a groundedness in the reality of the physical world. By this reading, to seek enlightenment apart from this world would be like looking for something that does not exist: a rabbit's horn.

In the Surangama Sutra the rabbit's horn is more complex. A beloved figure in Buddhism, Ananda was Buddha's cousin and one of his most devoted disciples. In the discourse related in this sutra, Ananda says that he has heard it proposed that the mind "exists nowhere and clings to nothing," which prompts him to ask if "that which does not cling to things" is, in fact, the mind. In answer to Ananda's question, Buddha replies, "You just said that the nature of the knowing and discriminating mind exists nowhere. Now in this world, all things in the air, in water and on the ground, including those that fly and walk, make the existing whole. By that which does not cling to anything, do you mean that it exists or not? If it 'is not,' it is just the hair of a tortoise or the horn of a hare." As in the Platform Sutra, it would seem that the horned rabbit, like the furred tortoise, functions as an allegory of that which does not exist. Also like the Platform Sutra text, it appears that in this passage our grounded world is asserted to be the true home of the mind. However, Buddha's point here is more nuanced, as he also indicates that nonattachment must exist as surely as does the physical world.

The Lankavatara Sutra, an ancient and central text in Mahayana Buddhism, contains a protracted discussion of the horned hare. Buddha begins this discourse by pointing out "there are some philosophers who are addicted to negativism," and who therefore "say that all things are nonexistent just like the hare's horns." In contrast to this view, Buddha continues, there are others who, because they are "attached to the notion that the hare's horns are nonexistent, assert that the bull has horns." Chastising philosophers for having "fallen into the dualistic way of thinking," Buddha proclaims that "the hare's horns neither are nor are not." Following an exasperatingly opaque discussion of the possible existence or nonexistence of the hare's horns, Buddha at last presents his interlocutor with this: "It is the same with the hare's horns, Mahāmati, whose nonexistence is asserted in reference to the bull's horns. But, Mahāmati, when the bull's horns are analysed to their minutest atoms, which in turn are further analysed, there is after all nothing to be known as atoms." Here, it would seem, Buddha is asserting not that the rabbit's horn does not exist, but rather that the bull's horn does not exist!

I am neither a student nor a practitioner of Buddhism, so the apparent contradictions within the Lankavatara Sutra—not to mention the incongruities between how the hare's horn is deployed across the three sutras—drove me to seek help. In addition to being a talented and inventive writer and photographer, my old friend Sean O'Grady has long been a practicing Buddhist, and when it comes to talking philosophy of any kind, he is among the most interesting people you might hope to meet. I sent Sean my notes, references to the passages in question, and an invitation to a conversation. We were soon on a video call and supplied with enough beer to try to grasp the horns—or, perhaps, the hornlessness—of the dilemma Buddha's teachings had presented me with.

I began by reading aloud the passage quoted above, from the Lankavatara Sutra.

> So, Sean, here's where I'm stuck. In the Platform and Surangama Sutras, it seems like Buddha offers his students the rabbit's horn as an image of something that doesn't exist. And he kind of says, "Get your head back into this world—the world of real things." But here, in this passage from the Lankavatara, he seems to complicate that idea. It's almost like he's saying "Hey, you say the rabbit's horn doesn't exist, while the bull's horn does. But that kind of dualistic thinking is a trap. How can you be sure the bull's horn really exists?" In saying the bull's horn only exists in our imagination or perception of it, is Buddha maybe saying that the rabbit's horn doesn't exist, but somehow the bull's horn doesn't exist either?

Sean's reply sought to shift my approach to the problem.

> I think you're on the right track, but the conclusion isn't that "nothing is real." It's just the impetus to go further. But to go further, you're not going to be using the logic that got you to this point. You have to leave that behind now. One of the famous metaphors in Buddhism is the raft. You get on the raft to get across the river, but once you get on the other side of the river, the raft is useless. Leave it! You don't need it! And this is just one river. What the Buddha is trying to get across—and all the koans [paradoxical anecdotes used as a teaching tool in Buddhism] are doing the same thing in their own way—is that you're

> applying a tool that doesn't work past a certain point. That tool in this case is logic. Logic is the raft.

I scrutinize Sean's face on the screen of my laptop. He's looking straight at me, with an expression that seems to suggest that he's saying the most obvious thing in the world.

> I mean, look, Sean, I'm just trying to figure this thing out. You know, this superficial reading that the hare's horn is simply an image of that which doesn't exist doesn't make a hell of a lot of sense in a world where the reality of the bull's horn is also in question. I mean, what's real if not the horn of a bull? Then again, as I sit here right now, I can't see the horn of a bull or the horn of a rabbit, though I can imagine them both. To me, in this moment, the two are equally real—they both exist. Do you think that might be what the Buddha is getting at?

Again Sean replied in a way that was intended to encourage rather than to chastise.

> If you keep trying to think your way out of it, you're going to fail. The horned rabbit, or the unicorn, or all of those figures that Alice meets in Wonderland, we in our culture say they're not real. They're just products of the imagination. And, from a certain perspective, that's true. But on an experiential level, they're quite real. They're just not the same kind of real. Because we're embodied, we actually live in the realm of the bull. That's where we always start from. But with our minds we can go to other places. This isn't a cult. Anybody who daydreams has been in that realm.

"Ok, that helps," I say. "So, maybe the metaphor of the rabbit's horn is supposed to help us understand that imagination is a legitimate form of reality—or, at least, that imagination can construct a legit reality. Maybe the Buddha means that a horn we dream of can somehow be as real as one we might be gored by? In any case, Buddha clearly valued the horned hare as a teaching tool, even if I can't be sure of what he meant to teach. I mean, bottom line: does the horned rabbit exist, or not?"

"Language always trips you up," Sean replied. "But when you inhabit the metaphor, when you imaginatively project yourself into it, then you experience it. Have you ever heard this familiar koan?

> Two monks were arguing about the temple flag waving in the wind.
> One said, "The flag moves."
> The other said, "The wind moves."
> They argued back and forth but couldn't agree.
> Hui-neng, the sixth patriarch, said, "Gentlemen! It is not the flag that moves.
> It is not the wind that moves. It is your mind that moves."
> The two monks were struck with awe.

Don't overthink this. You'll know your teacher when you see it. That teacher could be a human, or it could be a jackalope. I wouldn't preclude any possibilities."

Later that night, Sean sent me an email with a subject line that read only, "This ought to clear things up." Attached to his message was the fifty-fifth "case" in the Blue Cliff Record, a translation of the Chinese Ch'an Buddhist *Pi Yen Lu*, a collection of one hundred famous Zen *kung an*, or "public cases," koan-like stories that are supplemented with commentaries and verses from the teachings of Chinese Zen masters. Called "Tao Wu's Condolence Call," this "case" is augmented by the following verse by Sung dynasty Ch'an master Hsueu Tou Ch'ung Hsien (980–1052 C.E.):

> *Rabbits and horses have horns—*
> Chop them off. How extraordinary! How fresh and new!
> *Oxen and Rams have no horns.*
> Chop them off. What pattern is being formed? You may fool others.

In his metacommentary on the case, Yuan Wu K'e Ch'in (1063–1135 C.E.) asks, "Tell me, why do rabbits and horses have horns? Why then do oxen and rams have no horns?" Only when you understand this, declares Yuan Wu, can you understand the case. Yuan Wu continues,

> Some mistakenly say, "Not saying is saying; having no phrases is having phrases. Though rabbits and horses have no horns, yet Hsueh Tou says they have horns. Though oxen and rams have horns, nevertheless Hsueh Tou says they don't." But this has nothing to do with it. They are far from knowing that the Ancient's thousand changes and ten thousand transformations, which manifest such supernatural powers, were just to break up the ghost cave of your spirit. If you can penetrate through, it's not even worth using the word "understand."
>
> Rabbits and horses have horns—
> Oxen and rams have no horns.
> Nary a hair, nary a wisp—
> Like mountains, like peaks.
>
> These four lines are like the wish-fulfilling jewel. Hsueh Tou has spit it out whole right in front of you.

Only Sean would send along such a mind-bending text with a subject line indicating that it would provide total clarity. I sensed that he understood my deep desire to wrest control of the horned rabbit's mystery through my sheer tenacity as a researcher, and he wanted to provide a helpful challenge to my entrenched mindset—to "break up the ghost cave of [my] spirit." My friend had gifted me the vital reminder that the lure of the jackalope ultimately remains inextricable from its inscrutability. In his message accompanying this provocative, enigmatic text, Sean had written only a single line: "One will get quite far just reading and laughing one's way through koans."

Chapter 8
Dr. Shope's Warty Rabbits

The horned Cottontail is a well-known freak. The horns are a morbid product of the skin, and seem to be the result of irritating attacks by a skin parasite. The growths are of no particular service to the owner; and will hardly be interesting until someone more fully explains their origin.

—Ernest Thompson Seton,
Lives of Game Animals (1929)

The initial observation of the disease was made at Camp Funston, an Army training facility near Manhattan, Kansas, on March 4, 1918, when a mess cook arrived at the base infirmary presenting flu-like symptoms. By fall of that year, the worst pandemic in human history was ravaging every corner of the globe. As John M. Barry writes in *The Great Influenza*, the so-called Spanish flu "killed more people in a year than the Black Death of the Middle Ages killed in a century; it killed more people in twenty-four weeks than AIDS has killed in twenty-four years." The unprecedented virulence and lethality of the disease was visible even in the streets, where stricken victims sometimes fell prostrate, hemorrhaging fatally. "Their skin turned dark blue with the characteristic 'heliotrope cyanosis' caused by oxygen failure as their lungs filled with pus, and they gasped for breath from 'air-hunger,' like landed fish. Those who died quickly were the lucky ones. Others suffered projectile vomiting and explosive diarrhea, and died raving as their brains were starved of oxygen."

Unlike the COVID-19 pandemic, which in its early stages exacted a terrible toll on older people and those whose health was compromised, the 1918 H1N1 virus disproportionately struck younger, healthier people. The disease proved so severe among people ages fifteen to twenty-four that the outbreak may have killed 8 to 10 percent of young people then living, and almost overnight the life expectancy rate in the US plummeted by twelve years. Roughly 40,000 of the 100,000 American soldiers who died in World War I were killed by the virus, as troop movements spread the disease widely and intensified its effects by congregating the sickest soldiers into crowded field hospitals. The prevailing theory is that victims succumbed to a "cytokine storm," an inflammatory auto-immune response that severely damages the lungs. Younger, stronger people were hit hardest because it was not the virus itself but rather the body's immune response to it which often proved fatal.

Climate may also have played a role. From 1914–1919 the weather in Europe was unusually wet and cold, which exacerbated the spread and severity of the disease, especially among soldiers exposed to the elements while engaged in grueling trench warfare. This severe climate anomaly may have been caused by the war itself, as continual aerial bombardment filled the atmosphere with dust particles which served as cloud condensation nuclei, leading to increased precipitation. Like global climate change today—and also like the vast black clouds of the Dust Bowl during the 1930s—the climate irregularity that intensified the 1918 influenza pandemic was likely anthropogenic in origin.

The name "Spanish flu" is misleading. During World War I many countries, including the US, had strict media censorship policies in place that severely curtailed reporting on anything that might damage the morale of combat troops and the civilian infrastructure that supported them. Say, for example, a devastating, mysterious new disease epidemic. Because Spain remained neutral during the war, its press was free to report on the spread of the disease, which created an early association of the flu with that country. The virus would ultimately infect around 500 million people (then roughly a third of the world's population) and kill 50–100 million people worldwide, including an estimated 675,000 in the US. In a matter of months, the pandemic would take the lives of more people than any disease epidemic in human history. Despite its unprecedented lethality, in 1918 the cause of the world's worst disease outbreak was a medical mystery. It would remain so for fifteen years until Dr. Richard E. Shope and his rabbits finally provided an explanation.

Born on Christmas Day, 1901, Richard Edwin Shope was the son of a physician in Des Moines, Iowa, where his boyhood adventures included hunting, fishing, and exploring the wild lands surrounding his hometown. When he was ten years old, Shope began working on local farms, where he took an early interest in the health problems he observed in agricultural livestock. At age seventeen he went to nearby Iowa State College in Ames to register as a student in the School of Forestry but, finding the registrar's office closed, instead hopped a freight train and rode 130 miles east to Iowa City. There he enrolled as a University of Iowa premed student and, after two years of coursework, entered the university's medical school. Following his graduation in 1924, he remained for another year at the University of Iowa, where he taught pharmacology, conducted research on the chemotherapy of tuberculosis, and began to distinguish himself as a talented young scientist.

While an undergraduate at Iowa, Shope had been a member of the Student Army Training Corps (SATC, a precursor to today's Reserve Officers' Training Corps, or ROTC); the boys jokingly rendered the acronym SATC as the Saturday Afternoon Tea Club. When, in 1918, several of the young men in the Club died of influenza, Shope received an early and painful look at what many around the world would soon experience. The devastating global flu outbreak that followed would become an important part of Shope's fast-developing career as a research scientist.

The promise of Shope's early research came to the attention of established scientists and through the endorsement of his University of Iowa mentor, Dr. Oscar H. Plant, he was invited to join the Rockefeller Institute in Princeton, New Jersey. There he would collaborate with Paul A. Lewis, a viral pathologist celebrated for pioneering research leading to development of the polio vaccine. Shope had recently married fellow University of Iowa student Helen Ellis, and during the summer of 1925 the couple drove the Old Lincoln Highway from Iowa City to Princeton in a crank-start jalopy that had cost them only $35. Whimsically nicknaming their notoriously unreliable Model T "Galloping Asthma," the Shopes took the additional step of decorating the car with graffiti that editorialized on its obvious limitations, including "Don't Hit Me, I'm Getting Old," "Please Do Not Slam Doors," and "Shake Well Before Using." In addition to breaking down almost daily, the car had neither functional headlights nor floorboards, leaving the couple to ride with their feet on the dashboard. Remaining good humored despite the regular breakdowns, Shope wrote home to his parents on August 21, noting that "we certainly are getting the laughs along the road."

In an unpublished letter written from Princeton late in 1926, when he was just twenty-four years old, Shope explained to his parents why he had decided to remain at Rockefeller Institute and commit himself to a career in research rather than return home to take up his father's medical practice in Des Moines. "Now in doing medical research I am getting a shot at just the very things that have practitioners stumped at present and I can concentrate on one thing if I care to until I have either solved it or have to give it up for someone in a future generation (perhaps Dickie) to solve." (In fact, Shope's three sons, Richard Jr., Robert, and Thomas, would all go on to distinguished careers in viral studies.) "If I accomplish no more than to solve one single medical thing that is unknown now," the young Shope continued, "I would have accomplished much more than I would in a whole life sacrificed to general practice." As a son, I imagine the courage it must have taken for Shope to break the news to his family that he would not be following in his father's footsteps; as a father, I imagine how proud I would be of a child whose ambition was so devoted to the common good.

At the Rockefeller Institute, Shope collaborated productively with Lewis on important research into animal diseases. While studying hog cholera in the field, Shope witnessed his first occurrence of swine flu, which he and Lewis began to investigate in 1929. Working together to determine the cause of this devastating disease in swine, the men discovered and isolated a bacterium they called *Haemophilus influenzae suis*, which closely resembled *bacillus influenzae*, a bacterium that had been identified by German bacteriologist Richard Pfeiffer in 1892. At the time of Lewis and Shope's research, the so-called Pfeiffer's bacillus was widely considered to be the cause of human influenza. The hypothesis seemed credible; after all, bacteria had already been shown to cause a number of other diseases, including cholera and anthrax.

Soon after this new and promising avenue of inquiry was opened, however, Lewis succumbed to a risk inherent to virus research in its early years. Having traveled to Bahia, Brazil to study an outbreak of yellow fever there, he contracted the disease in his own laboratory and died within days. Shope continued the work alone, and in a series of influential articles published in the *Journal of Experimental Medicine* in 1931, documented his trailblazing research. Most critically, Shope had successfully transferred the agent of swine flu from sick pigs to healthy ones by filtering their virus-containing secretions. Filtration was key to early virus research because most bacteria are large enough to be removed by the porcelain laboratory filters then available. Viruses, on the other hand, can be more than fifty times smaller than bacteria and are too small to be observed with a conventional light microscope (this was before the commercial availability of electron microscopes starting in the mid-1960s); they slipped invisibly through the filter, thus prompting Shope's use of the term "filterable virus." The secret to swine flu, he found, was that the filterable virus he had discovered worked in concert with the bacillus he and Lewis had identified. In other words, Shope had discovered and isolated the extremely infectious swine flu virus, which he immediately suspected of being closely related to the human influenza virus.

In an unpublished letter written on January 18, 1932, Shope described his visit from Princeton to laboratories in New York, where he received from a fellow scientist several samples of human influenza bacillus. "I want to use them in my swine influenza work," he wrote, "to place as definitely as possible the relationship existing between the human and swine influenza." "A careful investigation would seem warranted," he would ultimately write in his published article, "of the possibility that Pfeiffer's bacillus and a filterable agent act in concert to cause influenza in man." Shope's speculation proved correct. Just two years later, in 1933, Christopher Andrewes and his research group at the National Institute for Medical Research in London succeeded in discovering the human influenza A virus, thus proving the human-swine viral link Shope had hypothesized. Andrewes, who became a lifelong friend of Shope's, would be knighted in recognition of his achievement.

In his subsequent research, Shope further demonstrated that the human and swine strains of influenza A (H1N1) were closely related by showing that serum derived from the human flu strain had the power to neutralize the swine flu virus. Having observed that the initial swine flu outbreak had occurred in 1918—timing suspiciously coincident with the outbreak of the devastating human influenza—Shope further speculated that the swine virus was in fact a surviving form of the human virus that had been responsible for the 1918 human pandemic. That is, he guessed that humans had passed the virus to pigs! To test his theory, Shope studied the antibodies present in people of different ages. In a 1936 experiment he documented that sera from children would not neutralize the swine flu virus, while sera from most older people would. Only those who had been alive to be exposed to the 1918 virus had developed natural protection against the swine flu virus, thus demonstrating the antigenic link between swine and human viruses. Richard Shope had at last identified the viral agent responsible for the most sweeping and lethal pandemic in history. To put this accomplishment in perspective, it was Shope's research that finally pointed to the cause of a pandemic that killed ten to twenty times more people than has COVID-19—a vast and incomprehensible loss that, until Shope's pioneering research, remained a terrifying medical mystery.

"I am guaranteeing that the name of Richard E. Shope will be remembered and spoken long after I am gone or anyone who knew me personally is gone," the young Shope had written in that remarkable 1926 letter to his parents. "If I do anything actually accurate and important my name will be attached to that fact as long as scientific literature is written or read and whenever that fact is quoted or used its origin will be stated." Although Shope had cracked the case of the worst pandemic in history, the work to which his name would be attached still lay ahead.

Shope's work on swine flu taught him a great deal about how viruses that ravage animals could shed light on—and even be related to—human disease. It was while conducting his pioneering research on swine flu that Shope became interested in tumors on cottontail rabbits (*Sylvilagus*), and it is here that the lines of Shope's story and that of the jackalope begin to angle toward convergence. Having returned to the Rockefeller Institute in October of 1931 from a research residency in Germany, Shope observed a tumor on the foot of a rabbit he shot while hunting near Princeton. By January 18 of the following year he was already noting, in a letter to his family back in Iowa, "I have had a bunch of box traps set up around the lab to get some wild rabbits for my tumor research." Shope soon discovered that the tumors were caused by another previously unknown virus, as he explained in that same January 18, 1932, letter: "It is a filterable virus disease that produces tumors which is brand, brand new. You see no tumor of mammals exists for which the causative agent is known. If it can be definitely proven that one mammalian tumor can be produced by a filterable virus then that will give a hint as to possibilities as to the causes of other mammalian tumors and might lead to a rational therapy of cancer." In 1932, the year after his breakthrough scientific paper on swine influenza, Shope published the results of his rabbit fibroma study. As his own letter suggests, he was already thinking about the possible ramifications of that work for cancer research more broadly.

As renowned virologist Peyton Rous explained in his presentation of the prestigious Association of American Physicians Kober Medal to his friend Richard Shope in 1957, Shope's discovery of the rabbit fibroma ultimately had enormous "human importance." In 1952, French bacteriologist Paul-Félix Armand-Delille, having heard of the effectiveness of the highly infectious disease myxomatosis in controlling rabbit population booms in Australia, deliberately introduced two laboratory-infected rabbits onto his French estate. Within six weeks, 98 percent of the rabbits on his property were dead. However, the virus soon escaped, and within a year approximately half of the wild rabbits in France—and a third of the domestic rabbits—had been wiped out. Ultimately, rabbit myxomatosis would spread across Europe. While the destruction of rabbits was welcomed by some farmers and foresters, it proved a serious blow to working people for whom the rabbit was an important food source. Fortunately, Shope had developed a vaccine; with the help of the Pasteur Institute, it was widely administered to domestic rabbits and over time helped to save the rabbitries of France.

Shope's rabbit fibroma research was only a gateway to the further study of virus-stricken rabbits, and it was his subsequent project that would lead to the greatest accomplishment of his career: his discovery and isolation of the rabbit virus that would come to be called Shope papillomavirus. In 1932, Shope became aware of stories of wild cottontails stricken with a disease that resulted in unusual growths on the animals that, he wrote, were "referred to popularly as 'horned' or 'warty' rabbits."

Like Richard Shope, virologist Ludwik Gross was a proponent of the theory that some cancers are caused by viruses. In his 1961 book *Oncogenic Viruses* (the first history of tumor virology), Gross included the following story, told in Shope's own words, reprinted with Shope's permission from his unpublished notes:

> The father of the wife of one of our staff members was visiting his daughter in Princeton shortly after I had started my experiments with the rabbit fibroma. This old gentleman was from Iowa and was quite a hunter out there. Because of this, his daughter had asked me if I would show her father the tumor that I had gotten from one of our New Jersey rabbits. I showed him what I considered a good example. He looked at it disdainfully and said that this was nothing compared with the sort of things that they had in the Iowa rabbits. He said he had shot rabbits with horns out of the side of their heads like Texas steers, or out of the top of their noses like a rhinoceros. Naturally, I was intrigued by this colorful description and speculated as to what the character of these growths might be.

When the hunter (T. A. McKichan, of Cherokee, Iowa) reported that such beasts were not uncommon in his home state, Shope traveled to back Iowa with him and spent several days hunting for horned rabbits. Failing to shoot any diseased cottontails, Shope initiated a contingency plan:

> We had a young boy by the name of Cliff Peck hunting with us, so when at the end of four days I had to return to Princeton empty-handed, I left a bottle of glycerol and a five dollar bill with Cliff and told him that if he would get me the horns of one of these rabbits and send them to me in glycerol, I would give him another five dollars. Needless to say, Cliff really scoured the underbrush for rabbits and within the course of a week I had my bottle of glycerol back and in it were several so-called horns from a cottontail rabbit shot near Cherokee.

Having studied the horns, sent to him in a 50 percent glycerol solution, Shope asked his contacts in Iowa and Kansas to procure him specimens of the anomalous rabbits if they could. A frantic search for diseased rabbits must have ensued, as a shipment of a dozen wild cottontails soon arrived in Princeton from Kansas, and before long Shope had received a total of seventy-five cottontails from the Midwest. Upon investigation, he discovered that eleven of those seventy-five animals were stricken with the same naturally occurring disease.

Shope now turned his full attention to the mystery of the "warty" rabbits. First, he devised a relatively simple experiment to test his unorthodox theory that the rabbits' odd "horns" were tumors caused by an unidentified virus. He removed and pulverized the horns, mixing the resulting material in a solution, which he strained through a porcelain filter so fine as to allow only viruses to pass through. If no virus was present in the resulting filtered solution, the solution would remain sterile. But when Shope rubbed the material that had passed through the filter onto the scarified skin of healthy rabbits—both cottontails and domestic rabbits—they subsequently grew horns! The results of this research on the diseased cottontails, published in his landmark 1933 article "Infectious Papillomatosis of Rabbits" published in *The Journal of Experimental Medicine*, demonstrated clearly that the grotesque excrescences on the cottontails were indeed caused by a previously unknown viral infection.

Shope had proven that the "horn" of the diseased rabbit resulted from infection by what would come to be called Shope papillomavirus. But he went further, documenting in detail the powerful neutralizing properties of the sera from infected rabbits. In a series of over one hundred wild and domestic rabbits which were experimentally administered the virus, not a single animal displayed natural immunity. Once rabbits were infected, however, they developed antibodies capable of resisting the virus. It was an immensely important finding. In the conclusion to his trailblazing 1933 article, Shope wrote that "Rabbits carrying experimentally produced papillomata are partially or completely immune to reinfection and, furthermore, their sera partially or completely neutralize the causative virus." Shope had not only discovered and isolated the virus that caused the growth of the rabbit's horns, he had also proven that the antibodies produced by stricken rabbits could effectively prevent reinfection. This latter discovery would open the door to new and even more exciting possibilities.

Shope's research on the warty rabbits suggested that the virus-induced growths were actually a keratinous carcinoma: a potentially invasive tumorous cancer. The importance of Shope's horned rabbit research was not simply that it revealed the viral etiology of the diseased animals' mysterious growths, but that it established a crucial link between viruses and cancer in a mammal. His related research on antibodies and resistance to viral reinfection hinted that the same virus that causes a certain type of cancer in a mammal might light the way toward development of an antiviral cancer vaccine. Then, as now, curing cancer was the holy grail of medical research.

And that is why what happened next is so surprising. Having published the monumental 1933 article on rabbit papillomavirus, and standing on the threshold of one of the most important breakthroughs in viral research, Shope simply handed the work over to his friend Francis Peyton Rous. More than twenty years Shope's senior, Rous was a colleague at the Rockefeller Institute, where he had arrived back in 1909, and where he later served as director of the institute's Laboratory for Cancer Research. Rous's early work on viruses in chickens resulted in the important discovery of what would come to be known as the Rous sarcoma virus—a retrovirus (a type of virus that inserts a copy of its RNA genome into the DNA of a host cell it invades) and the first oncovirus (a virus that causes cancer) to have been described by science. And now, thanks to Shope, Rous would have the opportunity to study a cancer-causing virus in a mammal.

In an unpublished letter to his parents, dated March 19, 1932, Shope hinted at his reason for giving away such promising research. "I have been very busy with the experiments I have in progress," he wrote. "Keeping the tumor & myxoma, and mad itch & influenza work all going at once takes lots of time and keeps me lumping." *Lumping* is defined by the *Oxford English Dictionary* as "to move heavily." In other words, Shope felt that the burden of running so many simultaneous projects was slowing him down. It is important to note that Shope usually conducted his labor-intensive research alone. In a posthumous appreciation, Colin M. MacLeod recalled Shope's belief that "the greatest joys and rewards of research come from carrying it out with one's own hands and mind—a trait which remained the mark of his own scientific way, commonly to the despair and amazement of his colleagues." That Shope simply had too many promising projects in progress was confirmed by fellow virologist Sir Christopher Andrewes, who wrote that Shope's work "gained a new dimension when it was found that in many tame rabbits the warts progressed and became carcinomatous. . . . Shope, at this time, was busy with many [research] problems, so he generously gave the material to Francis Peyton Rous. What Rous did with the rabbit cancers during the next thirty years is a matter of history."

Even more interesting is Peyton Rous's own account of how he came to receive the horned rabbit research from Richard Shope, an explanation included as part of the speech Rous gave in presenting the Kober Medal to Shope in 1957:

> The fourth new disease possessed by Shope in 1933, the infectious papillomatosis of cottontail rabbits, is under study wherever laboratories exist today for cancer research; yet he himself has done little with it. For in the very year of reporting his discovery he gave it over, with all its possibilities, to an [Rockefeller] Institute friend* in New York. He had shown by then that the virus produced growths much like tumors. . . . On a remembered day he came from Princeton and gave the virus to his friend to find out, giving indeed vastly more, all that the virus might mean for the cancer problem. . . . This the friend did and has studied its effects ever since as Shope's deputy.

If one follows the asterisk at the word "friend" to the bottom of the page in the transcript of the speech as published by the Association of American Physicians, Rous's note simply reads: "Namely myself."

Rous and Shope remained friends for life, and Shope was unfailingly appreciative in his comments about Rous's work. As late as June 3, 1965, when Shope delivered the Philip B. Price Lecture (under the title "Evolutionary Episodes in the Concept of Viral Oncogenesis") at the University of Utah College of Medicine, he noted that "The work that Rous and others have done in studying the malignant transformation of the rabbit papilloma has shown that it is a tumor that regularly progresses to cancer." That was, after all, the key finding to emerge from Shope's own unprecedented research on the horned rabbit.

The more deeply I studied the largely-untold story of Shope's brilliant research, the more obvious it became that real-life jackalopes had provided a crucial advance in our long quest to understand and prevent human cancer. My obsessive pursuit of the jackalope had led me to horned rabbits in nature, and the study of these virus-stricken rabbits brought me to the work of Richard Shope. Keeping my nose to the ground, I next discovered that the jackalope trail passing through Shope's life arrived at Nancy Helen Shope FitzGerrell and Thomas Charles Shope, the third and fourth of Shope's four children. Now in their mideighties, Nancy and Tom have keen memories of their father, and they kindly shared with me not only a trove of unpublished letters and photographs, but also many of their family stories.

Tom Shope had pursued a career not unrelated to his dad's, working for the Centers for Disease Control and Prevention (CDC), running experiments on Epstein-Barr virus in primates, becoming an academic at the University of Michigan, and eventually retiring in Ann Arbor. When I called Tom to get a better sense of his father, he was excited to share anecdotes about hunting, fishing, and gardening with his dad—passions Richard Shope pursued when he was not in his Rockefeller Institute laboratory. "He used to call me his Little Shadow," Tom said, with a warm chuckle. As a boy Tom had followed his father not only squirrel hunting and collecting specimens in the field, but also into his Princeton lab. "He would take me to his lab when I was just tall enough—I must have been four feet tall—just tall enough to see over the lab bench," Tom remembered. "I'd help him anesthetize mice. I'd watch him cut them open and take out the liver and the lungs." A future scientist himself, Tom even remembered seeing his dad performing some of his famous rabbit research. "I watched him take the domestic bunnies, the little white bunnies, and shave their bellies, and then take a little razor or knife and scarify their abdomens, and rub this material onto their skin, and then two weeks later there would be warts there. I remember seeing those experiments happen." When I asked Tom what he thought motivated his father, he replied that his dad's research "developed out of opportunities, and he had his ear to the ground and his eyes always open. He was very curious." Tom's appreciative characterization of his dad squared with my own sense that Shope's greatest power as a scientist was his inquisitiveness.

I asked Tom about his father's relationship with other scientists, especially those, like Christopher Andrewes and Peyton Rous, whose fame resulted from research directly based on his dad's work. "He had very, very close collegial relationships with people who were working on similar projects." According to Tom, Richard Shope and his colleagues "seemed to be very open with what they were trying to do, and they all seemed to have as a goal finding an answer to the problem rather than sequestering data so that they could use it later to advance their careers." Of Rous, the man who received Shope's warty rabbit research and followed it all the way to the Nobel Prize, Tom commented that "The relationship there was one of respect, and collegial sharing."

Then Tom provided a charming example. His storytelling skills on full display, he related an animated story of a visit by Christopher Andrewes to the Shopes' home in New Jersey.

> Sir Christopher came downstairs with his pants on and an undershirt, with suspenders hanging by his sides, hair disheveled, without his wire rimmed glasses, and said "Helen, something horrible has happened." Mom thought he had gone crazy by the way he looked and sounded. But then he explained that

> while pouring the water from the glass he had his false teeth in, he had accidentally flushed the toilet and the teeth fell through his hands into the toilet. He was scheduled to give a major speech in New York later in the day and we lived in Princeton, about two hours away on a good traffic day.
>
> We lived with a septic system, so Mom reached the septic system guy; he came out and emptied the tank, took the material to our back field to pour it out and look for the teeth, but didn't find them. He told Mom to flush the toilet again, while he held his hands at the pipe leading into the holding tank, and *voila*, there they were! Mom boiled his teeth after calling our dentist to be assured the teeth could be boiled. Christopher put them in, and he and Dad took off for New York City. He must have made it in time for the lecture, because that part of the story is vague. But the two of them reveled in telling the story for years afterward.

I had already been corresponding extensively with Nancy FitzGerrell, Richard and Helen Shope's only daughter, before I had the opportunity to talk with her. Now retired in Boulder, Nancy came out to Colorado to attend college and never went back. "My dad was very entertaining to me. He told stories, and we all laughed a lot. I just think I'm blessed with having a wonderful childhood," she told me. Her memories of home life featured her dad's love of the outdoors. "When he was at home, if it was spring, summer, or fall, he was outside working in the gardens, plowing the fields, making hay. We all helped put the hay up for the cow. The whole family pitched in." Like her brother Tom, Nancy had clear childhood memories of her dad's Princeton lab. "He would let us come in anytime. My mother dropped us off there sometimes. We could see his pigs in their cages, and his mice. I loved the smell of his lab," Nancy remembered. "He was very relaxed about people coming into his lab. It was a fun thing to do, to go and see him there." Her voice, already gentle, sounded girlish as she recounted what was clearly a treasured memory.

Nancy also recalled her father being deployed during World War II. "I remember when he went off to war, we had to take him to this train station, and everybody was pretty unhappy," she said, in a brave understatement that thinly veiled her sadness. World War II profoundly altered the trajectory of Shope's career. When war broke out, there was concern that the North American food supply might be vulnerable to biological attack and, in particular, that an infectious disease called rinderpest might be introduced into cattle. Called to service, Shope was appointed to direct a joint US-Canadian project to develop a rinderpest vaccine. Working with a team of five other scientists in a secret laboratory set up on a small island in the St. Lawrence River near Quebec, Shope worked tirelessly for a year and a half, ultimately producing an effective rinderpest vaccine that was deployed to safeguard the North American beef supply.

His next assignment proved an even greater challenge. A commander in the Navy, Shope joined an elite team of researchers whose charge was to determine what medical risks might threaten US and Allied soldiers engaged in offensives in the Pacific. When the Allies invaded Okinawa in a massive amphibious attack in early April 1945, Shope was a member of the medical research team that participated in the assault, setting up a field laboratory that at times came under fire in a battle so fierce that it resulted in more than 150,000 casualties. Nancy shared with me her father's letters to her mother while he was deployed—tender missives that are both personal and scientific. In my favorite of these, written at sea on April 8, 1945, Shope provides an indelible portrait of himself as the scientist about to enter battle, being lowered from a Navy ship into a small boat in preparation for landing. "I have my gas mask, rifle and my belt to which are attached two canteens, a knife, a jungle hat, a first aid kit & two clips of cartridges. In addition to this Lew Thomas & I have between us a cage of 56 mice & a bag of mouse feed. (The first mice to participate in an invasion, I believe)." (The Lew mentioned here is Lewis Thomas, who would go on to write award-winning books of popular science including *The Lives of a Cell* [1974] and *The Medusa and the Snail* [1979].) I admire Shope's bravery, his commitment to science, but also his aplomb in managing humor at what must have been a frightening moment. While in the Pacific, Shope continued his research, collecting several molds that he speculated might have scientific value. From one of these, a type of *Penicillium* he discovered growing on the isinglass frame of a photograph of his wife, Helen, he later derived an extract with powerful antiviral properties. It was later found that the potent new substance, which he named *helenine* in his wife's honor, stimulated the production of interferon, a signaling protein that is part of the body's natural defense against viral infection.

I asked Nancy (whose middle name, Helen, also honors her mother) the same question I had asked her brother about her father's relationships with other eminent scientists. "There was huge comradery," she replied, without hesitation. "I never got a sense that there was competition or worry about fame. They obviously shared stuff back and forth. And I think even in England the feeling was the same: What can we all do to make this work? All together. Not individually. Which is amazing." Nancy also speculated that the working conditions for her father's research were crucial to his success. "The way it was set up at the Rockefeller Institute was just pure science. They didn't even have to come up with a product like they would if they were in industry. They were all there to help one another." I asked Nancy what she thought motivated her father's work. "He wanted to feel that he had done something for mankind," she said. "He was definitely in the right job at the right time in the right place."

Although it remained unpublished until it appeared in the magazine *Veterinary Heritage* (the journal of the American Veterinary Medical History Society) in 2017, one of the key documents in understanding Shope's life is called, simply, "A Biographical Sketch of Richard E. Shope, MD." Dated May 12, 1952, this biographical essay is actually a high school term paper written by Nancy Shope when she was fifteen years old. The final lines of Nancy's class essay, submitted seventy years ago, speak volumes. "I have chosen R. E. Shope for my biography project because of his most interesting and eventful life. I only wish I had the time, effort, and skill to write a more complete biography of him, but right now I will leave that until some other time."

Shope's work exposed him to many dangerous viruses, and in the course of his research he was infected with two serious viral diseases, lymphocytic choriomeningitis and eastern equine encephalomyelitis, both of which he recovered from. His career as a medical researcher included the pivotal isolation of the swine flu virus and demonstration of its relationship to the global flu pandemic of 1918, the development of the rinderpest vaccine to protect the North American food supply, the heroism of his participation in the invasion of Okinawa during World War II, and his development of the anti-viral extract *helenine*. But no accomplishment in Shope's remarkable career would prove more important than his finding that papillomavirus was the cause of the strange growths on the heads of cottontail rabbits. Because Shope's discovery of the virus that causes rabbits to produce horns provided epidemiologists and oncologists with the first mammalian model of a virus-induced cancer, it ultimately led to crucial advances in the development of safe and highly effective antiviral cancer therapies.

As a world-renowned virus hunter, Richard Shope specialized in solving mysteries. But here is one mystery that remains unsolved. What connection, if any, exists between the hoax jackalope created by the Herrick brothers in Douglas, Wyoming, and the actual horned rabbits studied by Richard Shope in Princeton, New Jersey? Shope's answer to the question of what caused the 1918 flu contagion began with his simple, perceptive observation that the outbreak of swine flu and the epidemic of human flu occurred at the same time. Here is the coincidence that I have observed but am unable to account for: Richard Shope began studying horned rabbits in 1932, the same year that, according to Ralph Herrick, he and his brother Doug created the first jackalope mount in Douglas, Wyoming.

Douglas is 550 miles west of Cherokee, Iowa, the source of the first horned rabbit Shope ever examined. And yet, Mike Herrick, son of the jackalope's inventor, told me he had never heard of Richard Shope. Both Tom and Nancy Shope told me they could not recall their father ever speaking of the jackalope, though he devoted much of his professional life to the study of horned rabbits. We know that wild cottontails stricken with Shope papillomavirus were not uncommon throughout the Midwest; indeed, the prominent naturalist Ernest Thompson Seton called the horned rabbit a "well-known freak" as early as 1929. But we do not know the precise relationship between the hoax jackalope of lore and the virus-stricken horned rabbit of science. Perhaps that relationship is best left hidden in the jackalope's shadow. As Daniel S. Simberloff observed, "The jackalope may be an example of a new species produced by a virus acting in a completely different way—one in which the vector is imagination." Still, I wish I could sit down with the Herrick brothers and with Dr. Shope—perhaps to share a few rounds at the bar of the old LaBonte Hotel, in Douglas—and have them regale me with their tales of hunting rabbits, and viruses, and mysteries.

Chapter 9
Saved By Jackalopes

Tumors destroy man in an unique and appalling way, as flesh of his own flesh which has somehow been rendered proliferative, rampant, predatory, and ungovernable. They are the most concrete and formidable of human maladies, yet despite more than 70 years of experimental study they remain the least understood. . . . Few situations are more exasperating to the inquirer than to watch a tiny nodule form on a rabbit's skin at a spot from which the chemical agent inducing it has long since been gone, and to follow the nodule as it grows, and only too often becomes a destructive epidermal cancer. What can be the reason for these happenings?

—Peyton Rous, Nobel lecture, delivered December 13, 1966
("The Challenge to Man of the Neoplastic Cell")

Pathologist and virologist Peyton Rous had come to the Rockefeller Institute in 1909, when Richard Shope would have been just eight years old. The year after Rous's arrival, a farmer from Long Island brought him a Plymouth Rock hen with a substantial lump in her breast. After diagnosing the lump as a sarcoma (a malignant tumor of cells arising from connective tissue), Rous performed an experiment in which he extracted material from the tumor, strained it through a Berkefeld ultrafilter to remove bacteria and tumor cells, and then injected the filtered substance into a healthy chicken. The inoculated chicken subsequently developed cancer, suggesting to Rous that cells from the hen's tumor contained an infectious substance that was probably a virus.

Although Rous's seminal observation that a malignant tumor was transmissible would become foundational to modern oncology, at the time his research was met with skepticism and resistance by most fellow scientists. As Rous would later write, his discovery of what would become known as the Rous sarcoma virus was "met with downright disbelief." While Rous's case was hampered by the fact that no researcher had succeeded in demonstrating a link between a virus and cancer in a mammal, asserting the viral etiology of cancer in any species would have proved an uphill battle. At the time of Rous's early work, "Most scientists viewed cancer cells as basically anarchic—taking on a life of their own, entirely self-sufficient and unresponsive to normal cellular regulatory mechanisms." One of the chief objections to the idea that cancer could be infectious was the observation that medical professionals came into frequent contact with cancerous tissue and yet were not disproportionately vulnerable to the disease. Rous, the medical establishment maintained, was simply wrong. Those who attempted to discredit his findings were so certain of their position that "cancer researchers and physicians passed a formal resolution at an international conference in 1927 saying that 'it may be accepted for all practical purposes that cancer is not to be looked upon as contagious or infectious.'"

By the time that proclamation was made, Rous had long since abandoned cancer research. His inability to replicate his chicken sarcoma findings in mammals had proved an insurmountable obstacle to advancing the claim that viruses can cause cancer. The turning point in his career came when, as we have seen in the previous chapter, his Rockefeller Institute colleague, Richard Shope, gave his rabbit papilloma research to Rous. Shope's rabbit research showed not only that a virus could cause tumors—a finding that helped to validate Rous's chicken sarcoma work—but further demonstrated that it could do so in a mammal, thus bringing Rous's own work one important step closer to applicability to human cancers. Reprising the story of how he came to receive the rabbit papilloma project, Rous wrote that, while describing the growths on cottontails, "Shope remarked that they might be true [malignant] tumors. He knew of my fruitless search for a mammalian tumor virus, and he and I had long been friends. . . . Thus it came about that I experimented as his deputy throughout many later years."

In giving his warty rabbits to Rous, Shope had provided his colleague a clear path forward for research into virus-induced mammalian cancers. Rockefeller University summarized it this way: "Dr. Rous persisted in this line of research for the rest of his career and made discoveries that would determine the direction of the field for an entire generation of scientists." Working as Shope's "deputy," Rous's long and illustrious career would include a number of major breakthroughs in cancer research. He discovered that normal cells do not suddenly become malignant, as was then thought, but that their malignancy develops through discrete steps. He demonstrated that tumors are caused by viruses working in concert with other factors, including genes, hormones, and carcinogenic agents. He showed that the tumor cells of many species, including humans, can be grown in chick embryos, thus providing an important technique for research into the progression of cancer.

In recognition of his achievements as a cancer researcher, Rous received the Nobel Prize in Physiology or Medicine in 1966. The award came at age eighty-seven, more than a half century after his original research with the chicken sarcoma virus. That it took fifty-five years for his work to be fully vindicated (the longest "incubation period" in Nobel history) suggests the resistance that Richard Shope, Peyton Rous, and Christopher Andrewes encountered in proclaiming the viral etiology of some cancers. In an unpublished letter written by Andrewes to Rous on October 14, 1966, Sir Christopher congratulated Rous on his Nobel: "HURRAY! They should have done it several decades ago. What a triumph for righteousness and perceptiveness! . . . What a pity Dick [Shope] can't share in your pleasure." Peyton Rous's Nobel Prize was announced on October 13, 1966. Just eleven days earlier, his friend Richard Shope had died of metastatic pancreatic cancer.

Richard Shope called the virus he isolated from horned rabbits a "filterable virus." It was later named Shope papillomavirus (SPV) after its discoverer, later still renamed the cottontail rabbit papillomavirus (CRPV), yet later renamed again as *Sylvilagus floridanus* papillomavirus 1 (SfPV1), and ultimately recognized by the International Committee for Virus Taxonomy as Kappapapilolomavirus 2. What I will simply call Shope papillomavirus has proven to be a valuable model for the study of viral-host interactions, for the simple reason that research on virus-induced human cancers must proceed through the use of animal surrogates. Highly valued for its reproducibility in the laboratory, the Shope papillomavirus model has been widely used for decades to test various anti-tumor compounds. As Yale University virologist Janet L. Brandsma explains, "Animal models are essential to study the pathogenesis of papillomavirus infection and develop strategies for treatment and prevention. Because human papillomaviruses (HPVs) do not directly infect animals and there is no known papillomavirus for laboratory rodents, the only tractable animal model that enables study of papillomavirus infection and disease, including malignant progression, is the cottontail rabbit papillomavirus (CRPV)-laboratory rabbit model." The experimental importance of Shope papillomavirus was amplified in 1985, when Isabelle Giri and her colleagues at the Pasteur Institute in Paris published the complete genomic structure of the virus's nucleotide sequence. Dr. Shope's real-life jackalopes proved foundational to the scientific study of virus-induced cancers.

While on the trail of the jackalope, I tracked down Dr. Robert M. Timm, a wildlife biologist in the Department of Ecology and Evolutionary Biology at the University of Kansas, Lawrence. Timm, who may know more about rabbits infected with Shope papillomavirus than anyone in the world, grew up hunting cottontails in the Midwest. "Once in a while I'd come across a rabbit that had a disease and as a kid I didn't know what was going on, but was curious," he told me. "I suppose you could say I was a budding biologist from the start." He would go on to help build the university's collection of warty rabbits. "Because of my long-time interest in wildlife and in the effects of this disease, for several years I worked at obtaining specimens for the KU mammal research collection and there were some specimens in the collection already," he explained. "I was the curator in charge of the research collection. We have by far the largest collection of these rabbits in the world here as scientific study specimens. The papillomavirus causes quite weird horny growths on the head and neck, and in extensive cases elsewhere on the rabbit's body. No two rabbits infected with it look identical to each other."

Based on the data available to him, Shope believed his rabbit virus was confined to a portion of the Midwest, including parts of the Dakotas, Minnesota, Nebraska, Iowa, Kansas, Missouri, Oklahoma, and Arkansas. But Robert Timm, working with a team of molecular biologists, has profoundly expanded our understanding of the virus's range. Beginning with a physical examination of 1,395 rabbit specimens housed in the KU collection, Timm's research group first identified twenty individuals that appeared symptomatic. Using amplified DNA analysis to study those twenty specimens, the scientists were able to show that diseased rabbits came not only from the states Shope had identified, but also from Nevada, New Mexico, and Texas, as well as Jalisco and Nuevo Leon, Mexico. Further, they documented the presence of the virus not only in the Eastern Cottontail (*Sylvilagus floridanus*), but also in the Desert Cottontail (*Sylvilagus audubonii*) and, for the first time ever, the Mountain Cottontail (*Sylvilagus nuttallii*). (Earlier studies by other researchers had shown that the virus can also infect black-tailed jackrabbits, snowshoe hares, European rabbits, and brush rabbits, though it remains most common in cottontails—a fact that might help explain why jackalope mounts, despite the "jack" in their name, are sometimes fabricated using cottontails.) Timm and his colleagues were even able to extract DNA from a specimen collected in 1915, thus establishing the oldest lab-confirmed case of Shope papillomavirus—and a case that substantially pre-dates Shope's own pioneering research. The upshot of this fascinating study is that the virus that causes rabbits to grow horns occurs in a wider range of species than previously thought, and in a much wider geographic distribution than we ever knew. For decades, jackalope sightings and stories had circulated in lore from the desert Southwest and from Mexico, but it took a team of molecular biologists to prove that real-life horned rabbits do, in fact, inhabit those regions.

Papillomaviruses are members of a family of double-stranded DNA viruses called Papillomaviridae. As a nonenveloped (or "naked") virus, the papillomavirus has no lipid covering, and needs only its protein-based capsid and host detector proteins to invade host cells. More than 250 types of these viruses have been fully classified, and they are found to infect most mammals (including marine mammals), as well as birds, turtles, snakes, and (rarely) fish. Most papillomaviruses are host specific: they evolved over time to infect a unique host species, and are not usually transmitted between species. These viruses replicate only in the basal layer of body surface tissues. All papillomaviruses infect a specific body surface, most often the skin or thin tissue (the mucosal epithelium) around the genitals, anus, mouth, or throat.

That humans are not immune to the effects of papillomavirus is made clear in rare but dramatic cases in which people have grown cutaneous horns not unlike the growths on Dr. Shope's warty rabbits. In fact, the oddity of human horns has been documented for centuries. In the 1580s a Welsh woman, Margaret Gryffith, took the stage in London to exhibit the four-inch horn protruding from her forehead. It was said that Queen Elizabeth's Privy Council paused their preparations for the Spanish Armada to witness the spectacle. Driven out of his provincial village after having been accused of witchcraft, sixteenth-century Frenchman François Trouvillou also took to the stage, where he became a Parisian sensation for displaying the curved horn that protruded from his skull. Wang, the so-called "Human Unicorn," was a Chinese man who, in the 1930s, gained notoriety for the fourteen-inch-long growth sprouting from the back of his head. In his book *A Planet of Viruses*, Carl Zimmer relates the horrific story of Dede, a teenaged Indonesian boy who suffered from extreme growths caused by a papillomavirus. Sometimes called the "Tree Man,"

Dede suffered from growths that completely covered his hands and feet, causing them to resemble giant claws. In 2007, he underwent surgery and had an incredible thirteen pounds of warts removed from his body, though the warts continued to grow back, making subsequent surgeries necessary.

Phylogenetic analysis indicates that the basic genetic structural backbone of papillomaviruses may be more than 400 million years old, and there are more viruses out there than any other entity on earth. Even dinosaurs had warts caused by these viruses, and studies suggest that the earliest egg-laying terrestrial vertebrate, the common ancestor from which birds, reptiles, and mammals evolved, was host to papillomaviruses 300 million years ago. Genomics experts believe that "the established nucleotide variations observed in extant HPV genomes have been fixed through evolutionary processes prior to human population expansion and global dissemination." While papillomaviruses are much older than the human species, even human papillomaviruses are impressively numerous, prolific, and ancient. More than 200 HPV types have been genetically sequenced (and taxonomized into five genera), with more awaiting full sequencing. The most recent common ancestor of HPV is thought to be around forty to fifty million years old, which is astounding considering that, as a species, humans haven't been around longer than about six million years.

In fact, studies suggest that modern humans probably received some HPV strains from our not-quite-human ancestors. And this is the point at which sophisticated studies in molecular biology—studies with titles like "Transmission between Archaic and Modern Human Ancestors during the Evolution of the Oncogenic Human Papillomavirus 16"—reach popular culture in the form of articles with wonderful titles like "Yes, Humans and Neanderthals Had Sex. And They Gave Us an STD." Recent phylogenetic studies of HPV do suggest that early humans had sex with Neanderthals and another now-extinct, ancestral Paleolithic hominin species called Denisovans, and that these encounters were likely the means of papillomavirus transmission to *Homo sapiens*. Compare HPV's extraordinary pedigree to other viruses like measles or smallpox, which were passed from animals to humans around 1,500 years ago—or HIV, which made the leap only in the past century. HPVs, which may be the oldest of human viruses, are so ancient that they emerge from a shadowy time before we became fully human. We are fellow travelers, the virus having always been part of who we are as a species.

HPV, in all its glorious genetic diversity, is a sexually transmitted infection (STI), and that mechanism of transmission has been key to its tremendous evolutionary success. Because the fate of the virus is wedded to the human activity upon which the fate of our own species depends, it has been able to survive, thrive, and evolve along with us. (Though it is important to note that, because the virus is transmitted through direct skin-to-skin contact, it can also be spread through oral or anal sex.) The virus's means of replication are likewise impressive. HPV injects its own DNA into a human host cell, where, within the cell's nucleus, the host cell "reads" the virus's genetic code and creates the virus's proteins. Instead of killing the human host cells, as do many other viruses, HPV instead disables the cell's cancer-monitoring proteins, preventing them from sensing that the cell is being invaded. Then HPV hijacks the host's cellular mechanisms, tricking the host cell into making more copies of itself. This expertly choreographed dance between invader and host has been refined over the life of humans as a species.

In addition to being an ancient STI, HPV is widely distributed and extremely common. According to the CDC's analysis for 2018, one in five people in the US had an STI on any given day in 2018 (a total of sixty-eight million people), and during that year there were twenty-six million new infections, almost half of which were among people aged 15–24. The direct medical cost of treating those new infections added up to $16 billion for 2018 alone—a cost so exorbitant that it exceeds the total annual budget of a third of US states. Of the STIs tracked by the CDC (which include herpes, trichomoniasis, chlamydia, HIV, gonorrhea, syphilis, and hepatitis B), HPV was by far the most common, constituting almost two-thirds of total cases. The CDC expresses this prevalence bluntly: "HPV is so common that nearly all sexually active men and women get the virus at some point in their lives." The good news is that most HPV infections are benign, and the World Health Organization (WHO) reports that 90 percent of HPV infections resolve spontaneously within two years. Because many infections cause no symptoms, it is likely that you have had HPV without being aware of it.

Unfortunately, some HPV infections cause premalignant lesions or warts that can ultimately lead to cancer, although there is no way to know which infections will prove to be mild and which may ultimately develop into a potentially deadly disease. Equally vexing is the fact that HPV-caused cancer, when it does develop, has a long latency period, usually not appearing until ten to twenty years after the initial infection. HPV infections can lead to cancer of the cervix, vulva, vagina, penis, anus, mouth, tonsils, or throat. For example, it is estimated that HPV causes around 50 percent of penile cancers, 70 percent of vulvar cancers, and 80 percent of anal cancers. The cervical cancer statistics are truly startling: 99.7 percent of cervical cancers contain DNA from the so-called "high-risk" types of HPV. Of the more than 150 strains of HPV that live in the human body, relatively few are oncogenic (cancer-causing), among them types 16, 18, 31, 33, 35, 39, 45, 51, 52, 56, 58, 59, 66, and 68. Of these, types 16 and 18 together account for 70 to 75 percent of all cervical cancers.

Despite the resistance of the scientific community to Peyton Rous's findings, we now know that roughly 11 percent of cancer deaths worldwide are caused by viruses, which are the second most important risk factor for cancer, behind only tobacco use. Among these, HPV-caused cancers are a major cause of mortality. GLOBOCAN global cancer statistics estimate that, in 2020, 58,000 people died of cancers of the vagina, vulva, anus, and penis. Another 177,757 lost their lives to oral cavity cancers. While improved screening (Pap tests) in developed countries has substantially reduced the incidence of terminal cervical cancer, this disease—which, again, is caused almost exclusively by HPV—is still among the leading causes of death in women globally. In 2020 alone, cervical cancer killed 341,831 people. Even when the disease is not fatal, it can result in the need for hysterectomies, even in young women. And, as is too often the case, troubling inequities in health care remain a major factor. According to the WHO's *World Cancer Report 2020*: "The global disparity in cervical cancer incidence and mortality rates is an indicator of the enormous inequities in access to health services. Cervical cancer is the fourth most common cancer type diagnosed in women and the fourth most common cause of cancer death in women." They add that "Cervical cancer remains the most common cause of cancer death in many countries in Africa and South-East Asia, where the incidence and mortality rates are about 10 times those in North America." Inequities and disparities are also present in the US; for example, the cervical cancer survival rate is 20 percent higher for white women than for black women.

The cause of cervical cancer had long been a mystery, going back at least to the ancient Greek physician Hippocrates, who in 400 B.C.E. observed the disease and declared it incurable. Speculation about the etiology of this cancer ranged widely, including the far-fetched theory that cervical cancer was linked to eating bacon and ham. (An idea sparked by the fact that Jewish women have a low incidence of cervical cancer; however, the reasons for this are not cultural but instead genetic.) A more perceptive supposition was offered by mid-nineteenth-century Italian physician Rigoni-Stern, who speculated that the disease was sexually transmitted because he observed it to occur more often in married women and prostitutes than in virgins and nuns.

In a case made famous by Rebecca Skloot in her book *The Immortal Life of Henrietta Lacks*, Henrietta Lacks died of cervical cancer on October 4, 1951. The so-called HeLa cell line, derived from Lacks's cancerous cells, was propagated (unconscionably, without her knowledge or consent) in the laboratory and has been vital to cancer research. Today it is the oldest and most commonly used in vitro human cell line. The breakthrough research on cervical cancer was conducted, using that renowned HeLa cell line, by Harald zur Hausen, a German virologist whose work was the first to prove the connection between HPV and cervical cancer. When asked in an interview how he devised his theory, first published in 1976, that cervical cancer is caused by HPV, zur Hausen invoked two scientists whose work was directly connected to horned rabbits. "My hypothesis was based on several aspects," he replied. "In the 1930s, Richard E. Shope and Peyton Rous had previously demonstrated the potential carcinogenicity of a papillomavirus infection, the cottontail rabbit papillomavirus."

During the 1960s, many scientists believed that Herpes simplex type 2 was the cause of cervical cancer. After a string of fruitless attempts to demonstrate a connection between herpes and cervical cancer, zur Hausen instead turned his attention to HPV. He soon discovered that HPV is an extremely heterogeneous family of viruses, and that only certain types lead to cancer. In 1983, his team succeeded in isolating HPV-16, and the following year they identified HPV-18, the two strains that together account for about three quarters of all cervical cancers. More than three decades after Henrietta Lacks's death, Harald zur Hausen's work finally explained the cause of the aggressive cancer that took her life: HPV-18. For his research into the mechanisms of HPV-mediated carcinogenesis in cervical cancer, zur Hausen won the Nobel Prize in Physiology or Medicine in 2008.

When Harald zur Hausen identified the two strains of HPV that cause most cervical cancers, he recognized that the breakthrough might lead to an antiviral cancer vaccine, and he approached several pharmaceutical companies with the suggestion that they work to develop an inoculation that would protect against HPV-linked cancers. Incredibly, the companies performed a market analysis which, zur Hausen recalled, "indicated that there would be no market available" for such a vaccine.

Despite an initial lack of enthusiasm on the part of the pharmaceutical companies, the race to develop an HPV vaccine soon accelerated, as a number of research teams built upon zur Hausen's successful cloning of HPV-16 and HPV-18. The goal of each team was to trigger the body to produce antibodies that protect against HPV, and thus be the first to get a safe and effective vaccine across the finish line. The challenge, however, was to work out a way to activate the body's immune response without introducing carcinogenic viral DNA that would risk causing the cancer the vaccine was intended to prevent. This problem was solved in 1991 by Ian Frazer and Jian Zhou, at the University of Queensland in Australia, through the development of what are called "virus-like particles" (VLPs). Assembled through recombinant DNA technology, VLPs are hollow shells consisting of HPV coat proteins. The VLPs so closely resemble the natural virus that they prove to be strongly immunogenic: they trigger high levels of serum antibody production by the body, which responds as if it is being invaded by a real virus. The key, however, is that while the natural HPV virus capsid consists of two proteins (L1 and L2), the vaccine contains only one (L1). In other words, the vaccine contains no viral genetic material or live biological product, and so cannot be infectious.

Although the Australian scientists secured a US patent for their invention of the VLPs, research groups at the National Cancer Institute, Georgetown University, and the University of Rochester also made important progress on HPV vaccine development, and all four institutions filed patents between 1991 and 1993. In the years that followed, a number of pharmaceutical companies obtained partial rights to these patents. Legal battles ensued, leaving the companies unsure of who would ultimately control the patents necessary for vaccine production. After fourteen years of wrangling, the matter was settled in 2005 when pharmaceutical giants Merck and GlaxoSmithKline each signed cross-licensing royalty agreements with all four.

Soon after the patent dispute was resolved, two HPV vaccines became available in the US and thirty-four other countries. The first, marketed by Merck under the trade name Gardasil, was approved by the FDA in June 2006; in October 2009 GlaxoSmithKline received FDA approval for Cervarix. The HPV vaccine is now certified for use in over one hundred countries and is included on the WHO Model List of Essential Medicines, those drugs so safe and effective that they are deemed indispensable to any functional health care system. In 2018, sales of Gardasil alone topped $3 billion, and it has been estimated that in its first decade on the market the vaccine cut cervical cancer rates nearly in half. In November of 2020, WHO launched a global initiative to deploy vaccination, screening, and treatment to eradicate cervical cancer. WHO estimates that, by 2050, 40 percent of new cervical cancer cases and five million related deaths can be prevented through successful implementation of the proposed program. Adopted by 194 nations at the World Health Assembly, the resolution marks a new beginning to what we hope will be the end of the most lethal of HPV-caused cancers. In announcing the new global initiative, WHO Director-General Dr. Tedros Adhanom Ghebreyesus said, "Eliminating any cancer would have once seemed an impossible dream, but we now have the cost-effective, evidence-based tools to make that dream a reality."

In the fifteen years since its introduction, the HPV vaccine has been improved to respond to a greater range of carcinogenic strains of the virus. While the initial vaccines were bivalent (Cervarix, which protected against HPV types 16 and 18) and quadrivalent (Gardasil, which protected against HPV 6, 11, 16, and 18), the nine-valent Gardasil 9 vaccine, approved for use in December 2014 (and currently the only HPV vaccine on the US market), protects against HPV types 6, 11, 16, 18, 31, 33, 45, 52, and 58. This expanded coverage means that the vaccine provides protection against cervical, vulvar, vaginal, anal, and oropharyngeal cancers, as well as genital warts. Over time, the HPV vaccine has also been approved for expanded use. While the initial 2006 and 2009 Food and Drug Administration (FDA) approvals were for vaccine use in girls and young women (between the ages of nine and twenty-six years for Gardasil, and ten and twenty-five for Cervarix), in October 2011, the FDA expanded its recommendation to include boys aged 13–21 (though administration of the vaccine can begin as early as nine, and may be used in men up to twenty-six years old). In October 2018 there was a further expansion of the FDA's guidance, with use of the HPV vaccine being newly approved for men and women aged 27–45.

Despite this expansion, according to CDC guidance the ideal age for administration of the vaccine is 11–12, before children have been exposed to HPV strains that render the vaccine less effective. This is a crucial point: the vaccine should be given to kids at this age not *because* they are sexually active, but rather *before* they are sexually active. The vaccine has repeatedly proven safe and effective, and its use has been endorsed by the FDA, CDC, and WHO, as well as the American Academy of Pediatrics, the American Congress of Obstetricians and Gynecologists, the American Society of Clinical Oncology, the American Academy of Family Physicians, the American College of Physicians, and many other medical societies. Through a circuitous, intergenerational path of painstaking medical research, what ultimately emerged from Dr. Shope's work with warty rabbits is a vaccine so safe and effective that it has put the goal of ending HPV-caused cancers within our reach. There is a very real sense in which those who receive the HPV vaccine might well be saved by jackalopes.

Unfortunately, distribution and uptake of the vaccine has been much slower than its safety and efficacy warrant. One factor has been its high cost. The CDC reports that the average private sector cost per dose of Gardasil 9 is $239 (the recommended schedule, which varies by patient, requires either two or three doses). Despite the daunting price tag, in the US the vaccine is fully covered by most insurance plans and by the federally funded Vaccines for Children program for those who are eligible. More troubling has been the implications of this steep price for global distribution of the vaccine. A meta-analysis of HPV vaccine uptake published in 2021 indicates that in low- and middle-income countries only about 1 percent of women have received the full vaccine series. The coverage rate for boys is even more dismal. Indeed, most lower-income countries—where mortality from HPV-cased cancers is highest—have yet to include the vaccine in their national immunization programs. The authors of the study write, "Both the estimated uptake of the target population and the absolute number vaccinated represents the wide gap that needs to be bridged before achieving the WHO strategy of having 90 percent of girls fully vaccinated by fifteen years of age by 2030. The current estimate suggests that a significant proportion of women, especially in countries that started with low uptake of vaccination, remain largely unprotected and in settings that lack a proper cervical cancer screening program."

Even in the US, where the vaccine is widely available, uptake has been disappointingly slow. In fact, studies show that HPV vaccine has been the most often refused of common vaccines. Recent data published by the CDC indicate that even now, fifteen years after the vaccine was first approved for use, only about half of adolescents aged 13–17 in the US are up to date on their HPV vaccination.

To gain a better understanding of the reasons why so many parents remain hesitant to choose the HPV vaccine for their children, I called Dr. Elisa Sobo, Professor of Medical Anthropology at San Diego State University. Her research focuses on health, illness, and medicine, and she has done extensive work on pediatric vaccination and, more recently, on COVID-19 vaccine hesitancy. My first question was why a vaccine that is so safe and effective would meet so much resistance from parents. Sobo's insightful response focused on the psychological and social barriers posed by HPV being a STI:

> Here in the United States, the problem was that the vaccine was first marketed as a protection against sexually transmitted disease. And if you're an upstanding parent of a girl who is twelve years old, your child doesn't need that, right? There's a kind of denial, because to go out and get that vaccine is to say "my kid is having sex." So parents pushed back: "No child of mine," "My girl is a good girl," that kind of thing. It was just a big backfire. Some parents might even feel that by getting that vaccine for their kid, they're somehow endorsing them having

> sex. It should have been packaged as an anti-cancer vaccine, but it didn't come out that way.

Sobo has done fascinating studies on the concept of vaccine mistrust being socially cultivated: the idea that vaccine hesitancy within a community can become "contagious," spreading through social networks. I asked her if she thought this kind of social dynamic plays a role in resistance to the HPV vaccine. "Definitely it's a social thing, because you talk to other parents," she replied. "So they're talking to the other moms, saying 'Hey, have you got this for your kid?' Think about it this way: it's kind of like style, or fashion. All of a sudden, everybody has red shoes. Whether it's red shoes or vaccine hesitancy, there can just be a sense of 'this is what we do here.' It can become part of the community's identity." Then she added another interesting observation. "Also, with the HPV vaccine it's much less people talking to their elders for advice because it's newer. Nobody had this vaccine as a kid, so they can't draw on their elders' experience."

I asked Sobo if there are other reasons why parents might show vaccine hesitancy. "The problem with 'vaccine hesitancy' as a catch-all phrase is that it's like a big bucket into which we dump anybody who doesn't have uptake," she replied. Sobo proceeded to describe many different profiles of people who might be in that bucket: those who aren't able to get time off work; those who don't have insurance and thus can't afford preventative care; those who are in good health and so do not focus on the intangible threat of a cancer that might not develop for decades. "Then you have people who have other reasons to be suspicious," she added. "For example, a health care system that has systematically treated Latinos and African Americans unfairly. There's no real reason for them to trust the system, just based on historical fact." Finally, Sobo wanted to make clear that some forms of vaccine hesitancy result from a simple desire to gain a better understanding of the vaccine's risks and benefits. "Then there's this other part of the bucket, which is the people who just have questions, and they need the questions answered. But they're not met with respect. They're just tarred and feathered as 'anti-vax,' and so there are no answers to their very reasonable questions."

Given how many paths of resistance Elisa Sobo had identified, I began to better understand the challenges of communicating clearly about the HPV vaccine to people who might feel hesitant for very different reasons. As a way of comprehending what this problem looks like on the ground, I decided to ask my own family physician, Dr. John Hess, what challenges he faces as he goes about the important task of recommending the HPV vaccine to his patients. When I chatted Dr. Hess up during a routine physical, he told me he was aware of studies showing that the most effective point of contact for many vaccine hesitant people is their family doctor, and so he is keenly aware that he has a unique opportunity to help people understand the value of the HPV vaccine.

Dr. Hess told me that he sees HPV vaccine hesitancy often, and that he notices several different forms of resistance. However, to address this problem effectively he has to learn what his patients' concerns are, and he has found that there is no one-size-fits-all approach that works. "HPV-caused cancers are a little like tobacco. I can tell my patients that if they smoke now they might develop lung cancer in twenty years, but that's too abstract for a lot of people." Instead, he has developed a variety of practical techniques for conveying the message. "For example, when I'm talking with teenagers, I take a different approach with girls and boys. Girls are often more open to the cancer discussion. Teenage boys, that's a different story! With them, I focus instead on genital warts. I tell them I've seen huge, pustulous warts emerging from the tip of the penis. When I ask if they want a cauliflower on their penis, they're more open to the vaccine discussion."

Remembering Elisa Sobo's comment about the initial mismarketing of the vaccine, I asked Dr. Hess if he sees hesitancy linked to the fact that HPV is a STI. He answered that it is not uncommon for him to encounter resistance among deeply religious people who are committed to premarital sexual abstinence as a moral principle. "When talking to these folks I often make the point that some people may have strayed before they came into a committed relationship with the patient," he told me. "Even if my patient has never engaged in premarital sex, if their partner had other sexual partners before they were brought into the fold, that leaves my patient vulnerable. Most of us can relate to the idea that people sometimes make mistakes before they see the light, so explaining it this way really seems to connect with them."

When I commented that the HPV vaccine may be as close as we have ever come to discovering a cure for cancer, Hess paused before striking my left knee to test my reflexes. "It actually *is* a cure for cancer," he observed. "All of us are indoctrinated—even those of us in the medical professions—to the idea that a *cure* is what fixes something that's broken. But the ultimate cure is the thing that prevents disease in the first place." Given current debates about what constitutes "herd immunity" and whether it is practically achievable, I found Dr. Hess's final comment especially illuminating.

> HPV-related cancers are also a great reminder of why vaccine immunity can be superior to herd immunity. It's a little like COVID-19 in this respect. Yes, you might achieve herd immunity through transmission eventually, but you're going to lose a lot of people getting there. Most of us are exposed to HPV, and have been for a long time, but we now have a vaccine that can protect us against certain cancers. Why wouldn't we want to do that? Cervical cancer is real. I've lost patients to it.

In the fall of 2019, I was invited to do a reading of my work at the Colloquium on Mark Twain and the Environment, convened at Twain's beautiful, old home in Elmira, New York. While at the gathering I met Katherine Bishop, a talented scholar who was then Associate Professor of Literature at Miyazaki International College in Japan. As Katherine and I chatted about our projects, the subject of jackalopes and cancer came up, and she told me that when she returned home from the conference she would send me something I might find helpful. What she later shared has given me a new doorway through which to enter a vitally important conversation about HPV-caused cancers.

Katherine's stepsister, Julie Forward DeMay, was first diagnosed with stage IB cervical cancer in January 2008. The following month she underwent a radical hysterectomy (a serious surgery in which the entire uterus and cervix are removed, along with the top part of the vagina) in an attempt to remove the spreading cancer. It was hoped that all the malignant tissue had been excised, and Julie remained in remission until Christmas, at which time she was given much worse news. Despite the surgery, the cancer had metastasized, spreading to her chest wall and left lung. Julie's disease was now diagnosed as IVB, the most serious stage of cervical cancer, defined by the spread of the cancer to organs such as the liver, lungs, bones, or to distant lymph nodes. Because Julie had been thirty-four years old in 2006, when the HPV vaccine became available, she was outside the age range for which it was then approved and so did not have the option of protecting herself through vaccination.

The gift my new friend Katherine sent me was the journal Julie kept between New Year's Day 2009 and her death eight months later, on August 10. Wanting to share the story of Julie's journey and raise awareness about HPV-caused cancers and about the importance of the vaccine, in 2011 Julie's family posthumously published her journal as the book *Cell War Notebooks*. It tells a remarkable story, one that is extremely personal and deeply moving, but also beautifully animated by a spirit of resilience, courage, and humor. I think of the book as a rare record of how the mind and heart respond to an impossible situation in which the fate of the body can no longer be controlled.

Cell War Notebooks is punctuated by moments that are almost unbearably poignant: how Julie feels when she is no longer able to help her five-year-old daughter, Luka, do something as simple as pull on her tights; what it means to contemplate marriage vows exchanged with her husband, Scott, when neither of them could have imagined what loving "in sickness and in health" would mean when sickness becomes unimaginably invasive and destructive; or how it feels to lose your hair, as when Julie writes that "Being bald is the cancer signal and from here on out, I won't be so incognito with my damn ugly mutating cells. People will see my pretty scarf and know I am sick." At other moments, Julie reckons with larger, more philosophical questions rendered inescapable and urgent by her disease. "I continue to write in my journal for Luka," she writes. "While I continue to fight, I will live humbly alongside this unstable monster in my body. I do hope it doesn't erupt. But I do realize that, ultimately, it is a matter of that wondrous thing called fate."

Cell War Notebooks also brims with energy and hope. For example, here is the kind of rumination Julie shares as she imagines her relationship to her own cancer:

> Today I walked down to work in the sunny fresh air and imagined my little cancer cells going pop pop pop. I don't know if this is the sound it would make when they die, disintegrate, disappear, but it is how I imagine it sounds. As though the insides of my body can hear the little pops echoing, and are cheering and chanting wonderful shouts of victory. Imagery and visualization are supposed to help, so pop pop pop.

Although *Cell War Notebooks* is an intensely individual story written by a person with no time to spare, there is one moment in the text at which Julie reaches out to encourage readers to become better educated and more open about the risks of HPV-caused cancers. "Cervical cancer research has come a long way since the fifties," she writes. "Some people still don't like talking about it because of its association with HPV. But when close to 70% of the nation possibly are carrying the virus, with symptoms of its existence sometimes never surfacing, it seems logical to talk openly about it and to take serious measures to stop it."

Katherine also suggested that I talk with Julie's mother, Jane Forward, who generously agreed to share her experience with me in a conversation. When Jane answered my call, I was struck immediately by her unmistakable

Wisconsin accent, her friendliness, and her comfort with talking about Julie. "She was a very free spirit," Jane told me, her voice soft with affection. "She was very close in age with her siblings, but she always wanted to be her own person. She always used to say 'I don't want to be the other Forward, I just want to be my own person.' And she really was! She wasn't somebody who was in the box, and she was very feisty." Jane explained that Julie had studied creative writing at Colorado State University, and after graduation had lit out for new adventures in Portland, where at the time she knew no one. There Julie would meet the man who would become her husband, and give birth to her daughter, Luka.

"When she met Scott they fell in love, and they had a wedding on the beach. It was idyllic, and for two years everything was wonderful. And then, boom. Everything that could go wrong went wrong," Jane recalled. After Julie's initial diagnosis and hysterectomy, the family was hopeful. "She got better, and then she had that summer when she was doing really well. And we got together and rented a house on a lake and had a great time. Then the day before Christmas she went to the emergency room and found out it was back. And that's when she started the book. We had thought we were out of the woods, because the surgeon looked right at us and said she had gotten all the cancer. But she hadn't."

Since Julie's death, Jane has spoken at a number of seminars and schools, doing what she can to raise awareness about HPV-caused cancers and about the effectiveness of the vaccine. "Did you know about the link between HPV and cervical cancer before it touched your family?" I asked her. "No, I had no idea," Jane replied. "I didn't know anything about HPV. I didn't have the slightest idea about it until Julie was diagnosed. And Julie never told me that she got HPV. I found out about it after the fact, but she found out she had it when she got pregnant with Luka." Jane went on to explain that she has now learned much more about HPV, and that she hopes—both through *Cell War Notebooks* and through her direct efforts—to help others learn more too. "I spoke at a seminar for different kinds of cancers, and the man I sat next to was an EMT doctor, and he said the incidence of throat cancers in men has gone way up. And that's something a lot of people don't realize; it's not just cervical cancer, it's many different kinds of cancers that are caused by HPV. People think the vaccine is just for women, but it really isn't."

As Jane and I continued to talk, I had the pleasure of learning more about Julie, whose voice in her journal had so captivated me. How vivacious she was. Her creative work as a writer and photographer. Her deep love for her family and friends. I also learned more about Luka, who was just six years old when her mom died. Jane's pride as a grandmother came through when she reported that Luka had just graduated from high school and would be headed to the University of Oregon in the fall.

As we wrapped up our conversation, I asked Jane if there was anything more she wanted to add. "The only thing I want to say is that Julie was determined to beat it. She was just absolutely determined. And so when she was told there wasn't anything more they could do for her, that was a pretty rough time. It was about a month before she died when they told her that. That was a rough time for all of us." For the first time during our conversation, Jane's voice sounded unsteady.

Jane's moving story reminded me of the courageous passage that concludes *Cell War Notebooks*, the final words Julie was able to record in her journal: "It's just time to let go of the Pain. It's time to teach my daughter the beauty and strength in surrender; it's time to show her the absolute courage it takes to fight with all the power you have and then realize the Pain is not going to stop until you give it the word. . . . And when the Pain is gone, I can hear endless belly laughs on the porch and pretty music in the far off distance."

Chapter 10
The Jackalope Maker

Stuffed animals are cute—unless they once lived.

—Demetri Martin, standup comedy routine *Live (At the Time)*

When I was a boy, I bought a poorly mounted golden eagle from a junk shop in the town of Ruskin, Nebraska (population 123) while visiting my aunt and uncle there. Dark and cluttered, smelling of dust and leather, Deterding's general store had the feel of an old-fashioned, frontier mercantile. The place was crammed to the sills with everything a person might need—and plenty they would never need but might not be able to resist, like a stuffed eagle. I had no idea that the ten bucks I dropped on that ragged, beautiful bird violated the Eagle Protection Act of 1940, which had been expanded in 1962 to prevent trafficking in goldens as well as baldies. That disheveled mount presided over my bedroom throughout my childhood, and even on trips home from college I remember thinking that the bird seemed impossibly large.

Despite being the only kid in my neighborhood with his own stuffed eagle, I never took an interest in taxidermy. To begin with, you would have to kill an animal, and although I was an avid fisherman I somehow couldn't see my way clear to taking the life of a creature with fur or feathers. And beyond the killing was the preservation of the corpse, which seemed downright ghoulish. You might start out trying to earn the Boy Scout Merit Badge for Taxidermy (the scout was required to "present a satisfactory specimen of a bird or small mammal mounted by himself"), but who was to say you wouldn't end up like Norman Bates in Alfred Hitchcock's classic 1960 horror film *Psycho,* using your taxidermy skills to keep Mommy preserved in the cellar long after her soul had departed her body?

Visiting Mike Herrick's shop in Mills, Wyoming, however, had changed my view, because there I had witnessed the craft of taxidermy elevated to a fine art. As a result, it became clear that in my obsessive quest to understand the jackalope I had left the most obvious stone unturned: I had never attempted to make a jackalope myself. I was fortunate to have received a superb mount from Mike, and that meant a great deal to me. But if I was going to comprehend the jackalope inside as well as out, I needed to roll up my sleeves and have the unusual experience of becoming a jackalope maker.

It is more difficult than you might imagine to find someone who is both able and willing to teach you how to make a jackalope. Eventually I discovered Paxton Gate, a curiosity shop in the Mission District of San Francisco, not far from the old hippie stronghold of Haight-Ashbury. Trading in "curiously mind-expanding treasures and oddities," the shop also offers day-long taxidermy workshops, including one in which participants fabricate their own jackalope. The classes are expensive, and a quick reckoning suggested that I could buy three or four respectable mounts for what this trip to the city would cost. Nevertheless, the time had come to become a creator of the object I had so long sought to understand.

On the morning of the taxidermy class I walked five miles across San Francisco from my hotel but still arrived early at Paxton Gate, which was across the street from an Irish pub and a missionary Baptist church, and a little south of Good Vibrations, a store specializing in sex toys. An employee of Paxton Gate noticed me loitering on the sidewalk and invited me into the weird world of the shop, which was at once enchanting and slightly disturbing. From eye level up to the high ceilings, the walls were packed shoulder-to-shoulder with mounts of gazelle, springbok, wildebeest, elk, moose, caribou—even a giraffe with a graceful, six-foot-long neck, atop which balanced a beautifully tapered head, complete with glossy black eyes, cupped ears, and a pair of ossicones crowned with feathery tufts of black hair that made the giraffe look like it had matching sea anemones protruding from its skull. On the walls beneath the mounts were countless other natural wonders: sculls, teeth, the baculum (penis bone) of a red fox, beautiful feathers, collections of insects, wild assortments of moths and butterflies pressed beneath glass. The store was part natural science museum, part cabinet of curiosities. Everywhere the splendor of life was preserved in astonishing artifacts of death, immortalized in captivating monuments and fragments of perishing.

Soon other aspirational taxidermists arrived and we were guided to a makeshift workshop at the back of the store. Here two work tables were surrounded by ten folding chairs. On the table, laid out neatly in front of each chair, was a rabbit, positioned so as to appear midleap. Despite the lively pose of the bunny now before me, it was clear enough that its leaping days were through. The animal was neither jackrabbit nor wild cottontail, but instead a domestically raised "meat rabbit" that had been purchased at agricultural auction. In response to a question from one of the workshop participants, Geoff explained that, although these animals were raised to be eaten, it would not be advisable to make a meal of rabbits that would spend the full day at room temperature. I found it disconcerting that the animal's fur was a lovely black-and-white, resembling rabbits I had seen in pet stores—the sort of silky, sweet bunny you might see being eased into the arms of a giggling child in preparation for pictures on Easter morning.

Before I had the chance to wax sentimental, the instructor of our class, Geoff Vassallo, introduced himself as a professional taxidermist who has been stuffing animals since he was ten years old. Geoff runs his own taxidermy shop across the bay in Fremont, but drives into the city to teach these day-long workshops. And while he did not ask us to introduce ourselves, over the course of the day I would meet my fellow greenhorn taxidermists. At my table was Dave, his elderly father, and his two teenaged sons, all of whom are hunters; they wanted to have the shared experience of making jackalopes. Also at our table was James, a young man attending on his own. When I asked James what had brought him to the workshop, he replied, "I see roadkill all over the place and I just think to myself, 'You ought to figure out how to do something with that.' So I picked up a raccoon that had been run over and just tried on my own to figure out how to skin it. You know how sometimes you just have a feeling you'll be good at something even though you've never tried it? I did a pretty good job with that raccoon. Now I want to learn more."

At the adjoining table was a woman who teaches anatomy to homeschooled children, and who was interested in the workshop as an opportunity to navigate the rabbit's hidden maps of muscle, sinew, and bone. The woman sitting next to her had been given the class as a Christmas gift from a friend and, as she politely confessed to me during a break later in the day, "I don't know that this is something I would have chosen to do." Finally, a pair of young hipsters—a man and woman, well-pierced and tattooed—rounded out the group. The man was a musician who had apparently finished a late-night set not long before our morning class. I overheard enough to surmise that the couple had gone straight from his gig to waffles and beers before coming to the workshop. He was tall, with long black hair, a black T-shirt, and a black ball cap bearing the insignia of an Austin, Texas bar called Pour Choices. (On a subsequent trip to Austin, I discovered that Pour Choices is just a quarter mile from famed dive bar the Jackalope, also on Austin's legendary Dirty Sixth Street. And the Jackalope, fittingly, is just down the street from the Museum of the Weird.) The musician's partner, a young woman with a sparkling smile, was the funniest person in the group; she would keep things light even when our work demanded intense concentration.

Geoff briefly introduced the basics of the process that would occupy us for the next seven hours. "The idea is we're going to take the rabbit's pelt off over its head, just like pulling a sweater off over your head, inside out. At every step I'll show you what to do. I'll just move around to different people and demonstrate on their rabbit, and the rest of you can gather around and watch."

The first order of business, Geoff said, was to warm up with some "bunny yoga." The intent here is to restore flexibility to the animal's body before beginning work on it. We were directed to pick up our rabbits and bend their bodies gently in different directions. The animal felt so heavy—so very real—that I found this first contact unsettling. I could tell that a few of the other participants shared my discomfort, but the hunters at my table seemed perfectly comfortable, and humor broke the tension as people became more creative with their bunny yoga poses. One woman at the other table had her rabbit doing the cobra pose, while another helped her bunny to repeat a sun salutation.

Once our rabbits were limbered up, Geoff passed around a box and instructed us to take from it a single razor blade, the simple hand tool we would work with all day. No shaft, handle, or other apparatus—just a razor blade pinched between the thumb and pointer finger. Our first job was to make a perfectly symmetrical incision all the way around the torso of the animal, just below its shoulders. Following Geoff's instructions, I pinched my rabbit's loose skin with the fingers of one hand while making the cut with the other. This demanded patience, as the goal was to avoid slicing through the skin and accidentally piercing the membrane that holds the animal's viscera within its body cavity.

If this was a slow process, the next step was even more painstaking. The goal now was to pull the pelt gently toward the head, gradually and gingerly cutting the skin away from the body. If the cut was too deep, it would pierce the body cavity or, at least, leave too much muscle and meat on the pelt. If the cut was too shallow, the problem would be worse: piercing the skin meant that the jackalope, once completed, would have a noticeable hole in its furry hide.

This process continued for several hours, becoming more challenging as we approached the rabbit's face. Geoff coached us, demonstrating a technique that he made look simple, but that proved quite difficult, at least for me. The trick was to cut around the animal's eyes, ears, nose, and mouth without puncturing the skin. To do so, it was necessary to visualize the fine contours of the animal's face beneath the pelt, an act of imagination I found demanding. Because the inside-out skin was now pulled fully up over (but not yet off) the rabbit's head, it was impossible to see the face, whose delicate features were protected only by the thickness of the skin itself. One nick and the final project could be irreparably damaged. Geoff had a sixth sense for all of this, and he proceeded as if the skin were transparent, trimming gracefully around the facial features with surgical precision.

While much about the jackalope making process remained obscure to me, one thing was perfectly clear: I was the worst student in the class—a fact Geoff politely declined to contest when I volunteered the observation myself.

"Hey, Geoff, I think I might have cut too deep at the base of my bunny's ear," I said. He took a quick glance and said, "Yeah, you did! Take it easy around the ear. Get up for a minute." He took my seat and began working on my rabbit, muttering occasionally as he examined the damage my mistakes had already caused to my poor animal's pelt. "You've got to keep thinking about how this guy is going to look once you turn him right side out again," Geoff reminded me. As the process continued there was much discussion among the participants about what was beneath the skin at one spot or another. Was this the corner of the mouth? The base of the ear? The rim of the eye socket? Throughout this stage of the workshop everyone seemed both focused and cheerful, increasingly satisfied with their ability to perform the neat trick of turning a rabbit's pelt inside out. The exception was the female hipster, who, like me, routinely made mistakes. Unlike me, she didn't seem to mind at all.

Eventually the miniature hidden landscapes of our rabbits' faces were navigated more-or-less successfully, and Geoff demonstrated next how to cut through the hide around the wrists to allow the skin to release from around the rabbit's feet when the final step of removing the inside-out skin was performed. Once all necessary incisions were made, we would detach the rabbit's pelt from its body, a move made by holding the carcass with one hand and pulling the skin firmly away from it with the other.

When my rabbit's fur finally pulled free from its body, I was at first charmed. In my right hand I held a black-and-white bunny hand puppet: soft and smooth, complete with its cute, furry little face. But then I looked to my left hand, which held the animal's raw, skinless carcass. Its head appeared especially grisly, not at all recognizable as that of a rabbit. It looked more like the skull of a turkey vulture chick: hairless, pink, and wrinkled, with glossy black eyes bulging in a lifeless stare. To scan the work tables was to see this horror replicated, with our ten skinned bunnies strewn out across bloody sheets of newspaper.

For me, one of the hard but valuable lessons of the workshop was that a rabbit must die to bring a jackalope to life. That may sound obvious, but it is one thing to have a vague awareness of death at the margins of our consciousness, and another to face it directly. In fact, it now occurred to me, death surrounded the story of the horned rabbit. There was the perishing of the bunnies themselves, who had avoided death by coyote or by stewpot only to meet their fate at the hands of a taxidermist. There was the death of all those horned rabbits in the wild, whose papillomavirus-induced cancerous growths often made it impossible for them to eat, starving them to death. There was the death by dissection or experimentally induced cancer of the many rabbits in Richard Shope's Rockefeller Institute laboratory. Then there was the death of all the people who fall victim to fatal HPV-caused cancers—according to GLOBOCAN, 577,588 such deaths in 2020 alone. And there were the unthinkable numbers of people who will be killed by cancer in the coming decades, either because the HPV vaccine is inaccessible to them, or simply because they refuse it. But out of all that death also came life. If diseased rabbits had died in the bush and on Shope's lab bench, their deaths had also given birth to an extraordinarily safe and effective vaccine, one that has saved—and will continue to save—the lives of millions who would otherwise become the victims of fatal HPV-caused cancers.

None of these meditations slowed the process of jackalope making, which continued to be as enthralling as it was grisly. Our next step was to create the "form" over which the jackalope pelt would be stretched. This fascinating process involved laying the rabbit's skinless head on a piece of foam, carefully tracing its skull with a black Sharpie pen, and then marking the position of the eye by carving a shallow indentation. Next, we used a knife to sculpt the shape of the bunny's head, which was then made more precise by removing its rough edges by gently rubbing with sandpaper. Having skinned our rabbit and fabricated its form, it was almost time to break for lunch.

At just that moment I heard shouting from the front of the store. Craning my neck, I saw a woman standing just inside the doorway of the shop, holding a large sign whose message I could not make out, except to note that it included "PETA." "*IF NOT HERE, WHERE?*" she shouted. "*IF NOT NOW, WHEN?*" And with that she stepped outside without further confrontation. While it remained unclear whether the protestor's objection was to the store itself—full as it was of dead animals of every size and shape—or to the making of jackalopes specifically, I nevertheless felt a pang of guilt.

My fellow workshop participants responded to the commotion more gracefully than I did, though it was apparent that several of us were rattled, and in the moments that followed no word was exchanged about what had happened. For my own part, I felt torn. On the one hand, I was creating a charming piece of folk art that would bring joy to people—including my daughters, who always light up at the sight of a jackalope. On the other, I was about to dispose of the bloody carcass of an animal whose death I was responsible for, even though I had not killed it. Then again, weren't these meat rabbits, raised to be slaughtered? Yes, but raised to be slaughtered to be eaten, and there would be no rabbit stew simmering at the conclusion of our workshop. For that matter, was it better or worse to have killed a farm-raised rabbit rather than a wild rabbit? I wasn't sure. What about the ethics of food? It was undeniable that I had taken the rabbit's life without eating its meat, but at least I had used its pelt. And did it matter that I was turning its fur into folk art instead of making something practical, like a pair of slippers? The irony was rich, as San Francisco had recently passed an ordinance making it illegal to sell clothes made of fur. If instead of making a jackalope I had been making rabbit fur slippers, presumably I would be breaking the law.

It would take time to think through the ethical consequences of my choices—I wasn't ready to give up bacon, let alone jackalopes—but I appreciated that a well-intentioned stranger had provoked me into thinking more carefully about the complex ways in which any form of taxidermy both connects us to and alienates us from nature. Trophy hunters mount animals to commemorate their kill. Meat hunters eat the animals before putting their heads on a wall. Scientists use mounts for study and instruction. Jackalope makers employ taxidermy to bemuse and delight. But it was a sobering contemplation that, in all these cases, an animal still ended up dead. In bringing the dead to life, taxidermy also brought death to the living.

Having at last succeeded in skinning our rabbits, we sprinkled the inside of the pelts with borax, which serves as a drying agent and also contains mild antifungal and anti-insecticidal properties that help prevent rotting. We then placed our rabbit carcasses unceremoniously into a bin, cleaned up our workspace, and went our separate ways for lunch. I walked up Valencia Street to a Mexican restaurant, where I ordered vegetarian tacos and confessed to my journal the gruesome process necessary to create the whimsical jackalope.

Washing my tacos down with the last of a pint of Speakeasy Big Daddy IPA, I headed down Valencia Street to Paxton Gate. On the sidewalk in front of the shop I noticed one of the women from the workshop gazing into the front window, whose display included a full human skeleton, a taxidermied beaver, and a little stuffed white mouse mounted in an upright position on its rear legs, rocking joyfully on a tiny red swing.

"This place is a trip," I said, stepping up to get a better look at the whiskered, buck-toothed face of the beaver.

"Yes," the woman replied. "I love it here. This is my third time taking this workshop. People always have so much fun. Are you enjoying it?"

"I really am. Kind of gross, but totally fascinating. Why do you think people love jackalopes?"

She paused. "Well, I like the way making a jackalope combines nature and art. You have to really understand the rabbit's anatomy to do this right—there's precision required—but you also have to bring your imagination to the process."

Because I had arrived a bit early, I also had an opportunity to talk with Geoff, who was organizing some supplies at the back of the store. When I asked how he had become a taxidermist, he said that he grew up hunting and had taken an early interest in preserving animals.

"And how did you get into making jackalopes?" I asked.

Geoff said he had always enjoyed creating fanciful taxidermy sculptures. "Like when I mounted a horse for one of my clients and I added thirteen-foot-long wooden rockers to make it an actual, rideable, full-scale rocking horse. And check this out," Geoff said, extending his arm to show me a picture on his phone. In the photo was a relatively large primate mounted with raised, outstretched wings, posed aggressively with full-fanged snarl, perched on what looked like a small coffin.

"Is that a flying baboon?" I asked.

"I added the dyed turkey wings to an antique baboon mount, and I found that old chest used. Here's one more," he added, as the last of the workshop participants filed in to the work area. "I got a call from a guy who ran a small traveling zoo, who said his alligator had died. It was a seventy-five-year-old animal, eight hundred pounds, and had been very well cared for. My sons were teenagers then, and they helped me out. Together we skinned him and made this."

Geoff again extended his phone for me, though by now a small group of curious jackalope makers had huddled around to see the images of his incredible work. In this one, a fourteen-foot-long alligator was mounted on a pedestal, mouth agape with huge teeth gleaming. Its massive tail curved dramatically behind it, while its arms, feet, and even its long claws were extended as if to catch the wind. Its sides were graced with huge, outstretched black wings webbed dramatically with wildly veined lightning bolts of crimson. On its back rode a full-size leather saddle. Like the jackalope itself, the gator dragon was a magnificent hybrid creature, one born from nature's art and the fertile imagination of its creator.

The time had come to resume our work, so we returned to our stations and received Geoff's instructions. We now mounted the foam bunny head we had created on a prefabricated steel hanger by inserting the sharp point of the hanger into the form. The next move, among the most interesting of the day, was to turn the pelt inside out once again, and make its mouth "kiss" the mouth of the foam head. We then pulled the skin down over the form like a sheath, reversing the "sweater off" motion used in skinning the rabbit. If the alignment of the "kiss" is correct and the foam form has been shaped accurately, suddenly you are looking at your three-dimensional rabbit once again. All that remains is to tuck the extra skin of your jackalope sock puppet around the back of the hanger and pin it so the excess remains out of sight once the shoulder mount is displayed on a wall. At this late stage of the process I maintained my well-deserved reputation as the most inept student in the class, as my form was misshapen, and so the "sweater" of my rabbit's pelt did not fit quite right, producing an asymmetry that gave his slightly crooked face a sort of smart-ass smirk.

Among the final steps in jackalope making is placing glass eyes into the shallow sockets that have been created while marking the position of the eye. The idea is to press into place the gleaming eyes, which are dark brown with black pupils, and then pin the rabbit's fur around them. Done properly, no glue is required: as the pelt dries it will tighten, holding the eyes in place once the long, yellow-headed craft pins are removed. Although I used twice as many pins as did my classmates, my bunny's eyes repeatedly fell out of their sockets, bouncing off the table and onto the floor. Observing my unrelenting incompetence, Geoff set me up with a sort of putty he called "fire clay," which allowed me to reshape the poorly crafted sockets and also worked as an adhesive to help hold the rogue eyes in place.

The penultimate step was positioning my bunny's long ears. The goal is to situate the ears just so, and then pin them in place until the hide dries and tightens, after which the ears should remain in upright position. This is done by first tearing off eight bunny-ear-length pieces of cardboard, which are placed carefully, two each on the inside and outside of each ear, and then stapled together through the ears. Like the pins around the rabbit's eyes, the staples can be removed in a few weeks, leaving the lengthy ears charmingly perky. Or so I was told. My bunny, predictably, did not fare so well. During the earlier skinning process I had allowed a slip of the razor blade to cut through the rabbit's skin just at the base of the ear. The result was that my bunny's right ear simply flopped against its head, and no amount of cardboard and staples seemed to help. In desperation I again resorted to using a ridiculous number of pins, trying futilely to hold the ear upright. Despite my best efforts, my crooked-faced, limp-eared rabbit now looked like a stroke victim.

After nearly seven hours of bunny crafting, there remained but one final move, and that was to turn my rabbit into a jackalope by inserting small deer antlers into its head. Geoff had prepared the "horns" on nail-like spikes, so this stage consisted only of quickly piercing the rabbit's skin and pushing the spiked antler into the foam form of its head. Here, too, my technique was imperfect, and my jackalope's horns appeared a little lopsided.

As I continued to feel self-conscious about how poorly my jackalope had shaped up, I noticed that the woman hipster's creation had fared even worse. Unlike me, however, she was creative enough to regard every mistake as an opportunity. For example, when she cut through her rabbit's skin on its cheek, where the error could never be hidden, she turned the accidental hole into a third eye, so her jackalope looked like a rabbit from the right side and a double-eyed rabbit-flounder from the left. Then she doubled down on her jackafloundalope by purchasing two dried turkey claws from the store. Geoff quickly mounted them on spikes, and without hesitation the woman drove them into her animal's foam belly, thus producing a rabbit with horns, two eyes on one side of its head, and giant, scaly, outstretched talons. "He's radioactive!" she declared, expressing just the kind of delight a jackalope maker should always take in their work.

As I now stepped back to appraise what I had created over the course of a full day, my jackalope—whom I had named "Paxton" as a sort of appellation of origin (as if he were a fine cabernet)—seemed the perfect embodiment of my failure as a rookie craftsman. With his crooked face and flaccid ear and asymmetrical horns, his face full of numberless, yellow-headed pins—and even, I now observed for the first time, his slightly-crossed eyes—he appeared not so much poorly crafted as decidedly annoyed. I confess that his aggravated look charmed me, and the many frustrations I had experienced in making him now seemed preordained to produce a jackalope whose flaws were essential to his allure. I loved that Paxton seemed so irritated by my incompetence, and as he glared back at me with a look of genuine exasperation, I began for the first time that day to laugh. "Well, Paxton, you're absolutely right," I said aloud. "I sure fucked you up. No hard feelings?" I could almost see him rolling his crossed eyes as I imagined his reply. *Of all the people in the world who might have made me, I get this asshole.* I didn't blame Paxton for being disappointed in me, but, as I contemplated his endearing flaws, I could not have been more delighted with him.

By way of concluding the workshop, our group entered the small courtyard behind the store, where Geoff took a picture of us posing proudly with our newly minted jackalopes. It was a phenomenon I have seen repeated many times: perfect strangers spontaneously united by the seemingly universal appeal of the horned rabbit. As we smiled for the camera, I was struck by how joyful everyone was, regardless of the results of their efforts. I was joyful, too, because I would leave the shop having become something special that I had not been when I arrived: a jackalope maker.

I bid farewell to my fellow jackalopians and thanked Geoff for his patience. A guy working the counter at the shop kindly gave me an old cardboard box in which to transport my jackalope, such as he was, as I prepared to walk across the city. I placed Paxton carefully into the box, folded shut its flaps, and tucked the box under my arm.

Before leaving the store, I asked the guy if the jackalope-making class was popular.

"Oh, for sure. Probably our most popular. Fills up every time."

"Why do you think people love jackalopes?"

"It's cool in a weird way. Or, maybe weird in a cool way. But, I've seen lots of really different kinds of people in this class. Young, old. White, black. Teachers, hunters, artists. Neighbors who live here in the Mission. Probably different people have their own reasons for loving jackalopes. I guess all I can say for sure is that they do."

With my jackalope in tow, I stepped beneath the arcing neck of the giraffe and out into the low-angled afternoon light that poured into the city from across the Pacific. Leaving the Mission District, I sauntered north up Valencia Street, past thrift stores and art galleries, bars and used book stores, weaving my way through a thriving street scene that featured everything from poodles in pink sweaters to speeding electric skateboarders to the thumping bass of Arabic hip hop blasting out of a souped-up Honda and ricocheting off the brightly colored murals that blanketed the walls of alleyways. Walking through the city, I experienced a deep sense of satisfaction. What passerby could ever guess what marvelous treasure was concealed in my nondescript box?

I crossed beneath the steel girders of a highway and angled northeast onto the wider thoroughfare of Market Street, with its tall glassy office buildings, financial management firms, and upscale restaurants. Tracking the jackalope had been a wonderful journey, taking me to amazing places both real and imagined. I had been on the marvelous critter's trail for many years, doggedly following him to his remote haunts in history, folklore, literature, music, and the visual arts, and even pursuing him into the far fields of virology, oncology, and public health. I had made fascinating discoveries, finding traces of the horned rabbit in many world cultures, mythologies, and languages. I had met scores of eccentric, talented people along the way, from historians and folklorists, to taxidermists and collectors, to artists and musicians, to scientists and doctors. And my greatest pleasure had been to find, over and over, that while different people have different opinions about what makes the jackalope compelling, everyone seems to find the hybrid bunny a source of fascination, humor, or delight. In first descending into the horned rabbit hole, I could not have known the unseen worlds I would discover there. Now, as I walked with weird Paxton through the dying light of the city, I sensed that I was nearing the end of my long, strange, and wonderful trip.

Having ambled up Market Street for a few blocks, I hung a left on Polk, crossed the curving steel lines of streetcar tracks, and walked north again. On my left was the remarkable Beaux-Arts monument of City Hall, the golden fins of its immense dome illuminated in the low-angled rays of the setting sun. On my right was the open pavilion of Civic Center Plaza, where a throng of people were either celebrating or protesting something—though which it might be remained unclear from a distance. As I strolled with my horned bunny beneath my arm, I thought about the ancient traditions from which crooked-faced Paxton is descended. Indigenous peoples from around the globe have long told horned rabbit tales that teach us important lessons about nature, about each other, and about ourselves. In ancient times, Buddha had invoked the horned rabbit as a means of helping his students to ponder the very nature of reality. During the thirteenth-century, Abu Yahya Zakariya' ibn Muhammad al-Qazwini had described and depicted the venerable horned rabbit, Al-Mi'raj. From the sixteenth to the nineteenth centuries, European naturalists had attempted to understand and describe a presumed species, *Lepus cornutus*. During the twentieth century, the avatar of the horned rabbit that came to be celebrated as the jackalope captured the imagination of people dwelling in every part of the US. And I think of the jackalope's many global cousins, all those kauyumáris, ňwampfundlas, råkanins, skvaders, dilldapps, hasenbocks, rasselbocks, wolpertingers, oibadrischls, rammeschucksns, and raurakls to whom little Paxton is related. Even if you scored your jackalope at Wall Drug in South Dakota, or from Mike Herrick in Mills, Wyoming, or you crafted one yourself in the back of a weird shop in the Mission District of San Francisco, there remains an important sense in which your horned rabbit is a true voyager, world traveler, and global citizen.

And if the journey of the jackalope passes through Asia and Africa and Europe, it also passes through a small, prairie homestead outside of Douglas, Wyoming, where Doug and Ralph Herrick, in 1932 (according to Ralph), created the world's first jackalope. And that trail passes through a Princeton, New Jersey laboratory, where Richard Shope, also in 1932, used another kind of horned rabbit to show the world that a virus can cause cancer, vastly expanding our understanding of fatal human cancers and how they can be prevented. The Herricks shared their creation with the world, a genuine piece of folk art without patent or trademark that has brought joy to people everywhere. Shope gave his research freely to Peyton Rous, whose work ultimately ensured that Shope's legacy would include millions of lives saved from the brutal ravages of HPV-caused cancers.

As I walked through the Tenderloin, it was impossible to escape the disturbing contradictions so often apparent in the city. The sour smell of garbage rose from the gutters, and every side alley was packed with improvised tent villages inhabited by houseless folks. As I passed one of these ramshackle encampments, a Porsche raced past me, followed immediately by a Lamborghini. At last I crossed Geary Street and continued into the Polk Gulch neighborhood. Now several miles from where I had fabricated furry little Paxton, I was nearing the end of my remarkable journey. At last I arrived at the northeast corner of Polk and Post streets and walked through the door of my favorite San Francisco bar: Jackalope.

Taking a stool beneath the looming horns of the giant horned rabbit sculpture that graces this unusual tavern, I set my boxed treasure gently up on the bar. Next, I ordered their signature cocktail, the Jackalope, which is mixed with JP Wiser's rye whiskey, cherry Heering, Benedictine, and bitters, and garnished with lemon peel.

"Cheers," said the bartender, sliding the joint's eponymous cocktail toward me.

"Thanks," I said, in a slow pivot to my enduring question. "Why do you think people love jackalopes?"

"It's got to be the rye. That stuff is *fantastic*."

"Cheers," I said, lifting the glass to my lips and savoring that first, delightful sip.

"Whatcha got in there?" she asked. I looked at the nondescript cardboard box that lay between us, and then back at the barkeeper. Her friendliness and curiosity reminded me of the bartender I had spoken with in the bar of the LaBonte Hotel in Douglas, Wyoming, where my quest for the jackalope had begun.

"It's a long story," I answered, smiling.

As I sat quietly, sipping my delicious cocktail in the falling dark, I wondered where the trail I have been on for so long might lead in the decades—perhaps even the centuries—that lie ahead. The horned rabbit has undoubtedly had a glorious past, but what is the future of jackalopes? I don't suppose we can answer this question with any more certainty than we can know absolutely whether, as I believe, the jackalope of folklore was in fact inspired by the horned rabbits that have long hopped around the physical world. But the rich experiences I have had on the trail of the jackalope convince me that humans do need jackalopes. The horned rabbit's wild hybridity is captivating because it invites us into a strange species of joy that results from a blurring of the boundaries between the actual and the mythical. An elusive intercessor between real and imagined worlds, the jackalope is a trickster god who enriches both.

I am confident that the future of the horned rabbit is bright. It is our own curiosity and whimsy that brought the jackalope to life, and so only a failure of our collective imagination can bring about its death. In the enigmatic realm of the jackalope, the only possible cause for extinction is our failure to tell a story.

ACKNOWLEDGMENTS

Writers are very much in need of friends, and I have been fortunate to have so many in my corner. Here I offer my heartfelt thanks, along with equally sincere apologies to anyone I may have neglected to include.

At the heart of this book are the many conversations I've had with people across the country and around the world—folks whose lives and work have helped me to view the jackalope from every possible angle. Here, in alphabetical order, are the generous people with whom I conducted interviews for this book, or who offered important research assistance: Cecilia Åsberg (Guest Professor in Science and Technology Studies, Linköping University, Linköping, Sweden), Carrie Anderson Athay (Museum of Idaho, Idaho Falls, ID), Stacey Balkun (author of the poetry collection *Jackalope-Girl Learns to Speak*), Katherine Bishop (freelance editor, Los Angeles, CA), Anne Blaich (*Deutsches Jagd-und Fischereimuseum* [The German Hunting and Fishing Museum], Munich, Germany), Trevor Blank (Associate Professor of Communication and Interdisciplinary Studies, SUNY Potsdam), Alex Boese (creator, Museum of Hoaxes), Helga Bull (Douglas, WY Visitor Center), Alex CF (creator, Merrylin Cryptid Museum, London, England), Loren Coleman (director, International Cryptozoology Museum, Portland, ME), R. L. Crabb (comic book artist, Nevada City, CA), Joe DiLullo (American Philosophical Society, Philadelphia, PA), Emi (Jackalope Tattoo, Minneapolis, MN), Frank and Dianne English (taxidermists, Rapid City, SD), Linda Fabian (Wyoming State Historical Society, Wheatland, WY), Nancy Shope FitzGerrell (daughter of Richard E. Shope, Boulder, CO), Jane Forward (mother of Julie Forward DeMay, author of *Cell War*

Notebooks, Oconomowoc, WI), Mel Glover (Wyoming Pioneer Museum, Douglas, WY), Lyndsey Green (artist, Knutsford, England), Jim Herrick (taxidermist, Big Horn Taxidermy, Douglas, WY), Luke Herrick (Douglas, WY), Mike Herrick (taxidermist, Antler Taxidermy and Arts, Casper, WY), Dr. John Hess (physician, Reno, NV), Lee R. Hiltzik (Senior Research Associate, Rockefeller Archive Center, Sleepy Hollow, NY), Rick and Sarah Hustead (Wall Drug, Wall, SD), Christoph Irmscher (Provost Professor of English and George F. Getz Jr. Professor in the Wells Scholars Program, Indiana University), Bruce and Cathy Jones (Mayor of Douglas, WY), Bill Kalar (Hotel LaBonte, Douglas, WY), Lauren LaFauci (Assistant Professor of Environmental Humanities, Linköping University, Linköping, Sweden), Carrie Lambert-Beatty (Professor of Art, Film, and Visual Studies, and of History of Art and Architecture, Harvard University), Cynthia Mackey (Peabody Museum of Archaeology and Ethnology, Harvard University), Joe Mailander (musician, Okee Dokee Brothers band), Bob McGowan (Senior Curator of Birds, National Museum of Scotland, Edinburgh, Scotland), John Meissner (Director, Estes Park Archives and Sanborn Research Center, Estes Park, CO), Macey Moore (Hotel LaBonte, Douglas, WY), Logan Murphy (Paxton Gate, San Francisco, CA), Bill Nelson (artist, Denver, CO), Sean O'Brien (jackalope postcard collector, Layton, UT), Sean O'Grady (writer and photographer, Maplecrest-in-the-Catskills, NY), Lynda Olman (Professor of English, University of Nevada, Reno, Reno, NV), Elliott Oring (Professor Emeritus of Anthropology, California State University, Los Angeles, CA), Greg Ornay (Biodiversity Institute and Natural History Museum, University of Kansas, Lawrence, KS), Henrik Otterberg (Thoreau scholar, Gothenburg, Sweden), Chet Phillips (artist, Austin, TX), Michael Pinckney (archivist, Ronald Reagan Presidential Library, Simi Valley, CA), Cindy Porter (Chamber of Commerce, Douglas, WY), Sankurie Pye (Curator of Invertebrate Biology, National Museum of Scotland, Edinburgh, Scotland), Richard Reibel (creator, The Jackalope Conspiracy website), Andy Robbins (author of *Field Guide to the North American Jackalope*), Dr. Al Ruenes (urologist and surgeon,

Doylestown, PA), Jennifer Schmaus (*Deutsches Jagd-und Fischereimuseum* [The German Hunting and Fishing Museum], Munich, Germany), Kari Serrao (artist, Toronto, Canada), Julie Seton (Earnest Thompson Seton Institute, Las Cruces, NM), Thomas Shope (son of Richard E. Shope, Ann Arbor, MI), Elisa Sobo (Professor of Medical Anthropology, San Diego State University, San Diego, CA), Reena Spansail (artist, Reno, NV), Jenna Thorburn (Wyoming Pioneer Museum, Douglas, WY), Bob Timm (Department of Ecology and Evolutionary Biology and Natural History Museum, University of Kansas, Lawrence), Bill Truitt (Douglas, WY Visitor Center), Geoff Vassallo (Wilderness Taxidermy, Fremont, CA), Dave Watkins (filmmaker, Jasper County, GA), Andrew Williams (artist, Orlando, FL), Jen Womack (Chamber of Commerce, Douglas, WY), Hannah Yata (artist, Lords Valley, PA), and Heather York (Research Associate, American Museum of Natural History).

Among fellow nonfiction writers, thanks to Rick Bass, Paul Bogard, Taylor Brorby, Bruce Byers, John Calderazzo, SueEllen Campbell, Craig Childs, Laird Christensen, Chris Cokinos, Elizabeth Dodd, Tom Fate, Andy Furman, Amy Hammer, Dimitri Keriotis, Drew Lanham, Paul Lindholdt, Ian Marshall, Tracy Melton, Kate Miles, Kathy Moore, John Murray, Gary Nabhan, Nick Neely, Tim Palmer, Jodi Peterson, Bob Pyle, David Quammen, Eve Quesnel, Brad Rassler, Suzanne Roberts, Don Snow, Gary Snyder, John Tallmadge, David Taylor, Leath Tonino, Nick Triolo, and Rick Van Noy. Special thanks to John Price, John Lane, and David Gessner, whose support has been decisive.

Closer to home, thanks to fellow Great Basin writers Kendra Atleework, Alicia Barber, Caleb Cage, Chris Coake, Bill Fox, Shaun Griffin, Dave Lee, Mark Maynard, Gailmarie Pahmeier, Ann Ronald, Rebecca Solnit, Jared Stanley, Steve Trimble, Claire Watkins, Terry Tempest Williams, and Lindsay Wilson. I've also received valuable support from many Reno, Truckee, and Tahoe friends, among them Pete Barbieri, Fil Corbitt, Dondo Darue, David Fenimore, Cheryll and Steve Glotfelty, Tony Marek, Ashley Marshall, Ben Mathews, Eric Rasmussen, and Monica and Colin Robertson.

Thanks also for the encouragement I've received from friends in the environmental humanities and humor writing communities, including Tom Bailey, Christina Barr, Megan Berner, Jim Bishop, Chip Blake, Jenn Brown, Simmons Buntin, Ben Click, Nancy Cook, David Cremean, Ann Fisher-Wirth, Stephanie Gibson, Charles Goodrich, K. J. Grow, Tom Hallock, Tom Hillard, Josh Hohn, Heather Houser, Richard Hunt, Dave Johnson, Rochelle Johnson, Talley Kayser, Claire Kelley, Kathleen Kuo, Mark Long, Tom Lynch, Kyhl Lyndgaard, Rona Marech, Annie Merrill, Clint Mohs, David B. Morris, Henrik Otterberg, Phil Pacheco, Dan Philippon, Anna Lena Phillips, Justin Race, Lisa Rossi, Rowland Russell, Jennifer Sahn, Brian Schott, Heidi Scott, Robert Sickels, David Taylor, George F. Thompson, Jim Warren, Alan Weltzein, Tracy Wuster, and Boyd Zenner.

I'd like to acknowledge the Ellen Meloy Desert Writers Fund, the University of Nevada, Reno Foundation, and the Nevada Writers Hall of Fame for helping to support work on this project. Key to my research has been the expert assistance of a talented team of librarians, including Kathy Anderson, Kim Anderson, Donnie Curtis, Daniel Fergus, Georgia Grundy, Mark Lukas, Ann Medaille, Usha Mehta, Robin Monteith, Molly Mott, Michelle Rebaleati, Maggie Ressel, Amy Shannon, Teresa Schultz, Jacque Sundstrand, Kyle Weerheim, Jennifer Wykoff, and the gifted, patient, and generous Roz Bucy, who has been the horned rabbit's nerdy guardian angel. This project has received helpful research support from Dave Stentiford, Ashley Garver, and Megan Cannella. Very special thanks to jackalope trackers extraordinaire, Phoebe Wagner and Katie Wolf, whose tireless research helped to make this book possible.

I'm deeply grateful to my agent, Laurie Abkemeier of DeFiore and Company in New York, who not only landed this book amid the chaos of a global pandemic, but has expertly guided this stage of my writing career. Thanks to the team at Pegasus Books in New York: Derek Thornton (cover design), Maria Fernandez (production), Jenny Rossberg (publicity), Victoria Flickinger (copyediting), and Mike Richards (proofreading). Special thanks to Deputy Publisher Jessica Case. I couldn't be happier that my horned rabbit found a home in the stable of the winged horse.

I am blessed with a family that is exceptionally tolerant of my obsessions and idiosyncrasies. On the California side of the Sierra, thanks to our Central Valley people: O. B. and Deb Hoagland, Sister Kate and Adam Myers, Troy and Scott Allen, and the whole brood of cousins. Here on the Nevada side, thanks always to my parents, Stu and Sharon Branch, who have directly or indirectly enabled everything I have accomplished in life. I often tell our daughters that "it takes a family to make a book," and I am lucky to have such a supportive and encouraging crew. Thanks to Hannah and Caroline for rolling with the idea that their dad's obsession with horned rabbits isn't actually that weird. The dedication of a book is the sincerest gesture of gratitude available to a writer, and I consider myself incredibly fortunate to be able to dedicate this one to my wife, Eryn. I'm quite sure that when she agreed to marry me more than twenty years ago, she had no idea that jackalopes were in her future.

ENDNOTES

EPIGRAPHS

p. ix I don't see what's so impossible about: Anton Chekhov, letter to Alexei Suvorin (from Moscow, 11 Sept. 1888), *Letters of Anton Chekhov*, edited by Simon Karlinsky, translated by Michael Henry Heim (New York: Harper & Row, 1973), 107. This line refers to Chekhov's desire to pursue careers in both medicine and literature, and is an allusion to a proverb from Erasmus's *Adages*: *duos insequens lepores, neutrum capit* ("he who chases two hares catches neither").

p. ix "Hallo, Rabbit," he said, "is that you?": A. A. Milne, *The Complete Tales & Poems of Winnie-the-Pooh* (New York: Dutton, 1996), 112.

p. ix Ideas are like rabbits: John Steinbeck, *Conversations with John Steinbeck*, ed. Thomas Fensch (Jackson, MS: University Press of Mississippi, 1988), 43.

p. ix The truth is a rabbit in a bramble patch: Pete Seeger, interview in *Songwriters on Songwriting*, edited by Paul Zollo (Boston: Da Capo Press, 2003), 8.

p. ix People's dreams are made out of: Barbara Kingsolver, *Animal Dreams* (New York: HarperPerennial, 1990), 334.

AUTHOR'S NOTE: DOWN THE RABBIT HOLE

p. xiii "One *can't* believe impossible things": Lewis Carroll, *Through the Looking-Glass, and What Alice Found There*, in *The Annotated Alice: Alice's Adventures in Wonderland and Through the Looking Glass* (New York: Clarkson N. Potter, 1960), 251.

PROLOGUE: NATURE'S JACKALOPES

p. xv By 1932, Dr. Richard E. Shope: For more extensive references on Shope, see notes to Chapter 8 of this book, "Dr. Shope's Warty Rabbits."

p. xv "were said to have horn-like protuberances": Richard E. Shope, "Infectious Papillomatosis of Rabbits," *Journal of Experimental Medicine* 58, no. 5 (1933), 607.

p. xvi "black or grayish black in color": Shope, "Infectious Papillomatosis of Rabbits," 608.

p. xvi "white or pinkish white fleshy center": Ibid.

p. xvi "The first detectable lesions": Ibid., 609.

p. xvii "acquired a definitely warty appearance": Ibid.

CHAPTER 1: AS REAL AS YOU WANT THEM TO BE

p. 1 "'Monsters' can help us": Boria Sax, *Imaginary Animals: The Monstrous, the Wondrous, and the Human* (London: Reaktion Books, 2013), 249-50.

p. 1 The town of Douglas: For more on the history of Douglas, Wyoming, see Arlene Ekland-Earnst, Linda Graves Fabian, and Carol Price Tripp, *Douglas* (Charleston, SC: Arcadia Publishing Library Editions, 2010).

p. 2 class offered by the Northwestern School of Taxidermy: Materials from this correspondence course may be found at "Northwestern School of Taxidermy (Omaha, Neb.) [RG5451.AM]," *History Nebraska*, https://history.nebraska.gov/collections/northwestern-school-taxidermy-omaha-neb-rg5451am.

p. 3 "Let's Mount That Thing!": Douglas Martin, "Douglas Herrick, 82, Dies; Father of West's Jackalope," *New York Times* (January 19, 2003), https://www.nytimes.com/2003/01/19/us/douglas-herrick-82-dies-father-of-west-s-jackalope.html.

p. 3 Herricks are making jackalopes in Wyoming: For more on the Herrick family and jackalope taxidermy, see Chapter 4 of this book, "Mounting Enthusiasm."

p. 5 I slowly navigate the wide, quiet streets of Douglas: My first visit to Douglas was in October 2008. The interviews recounted in this chapter took place on October 17–19, 2018.

p. 12 An official proclamation: For the full text of this proclamation, see https://www.cityofdouglas.org/DocumentCenter/View/1038/Jackalope-Proclamation.

p. 15 "Here's our brand new logo": The Douglas logo with the stylized jackalope may be seen on the homepage of the town's website, at https://www.cityofdouglas.org/.

p. 16 the town's visitor center: For more information about the Douglas, Wyoming Railroad Museum and Visitor's Center ("Locomotive Park"), see https://www.cityofdouglas.org/Facilities/Facility/Details/Locomotive-Park-4.

p. 18 Jackalope Days, a festival the town holds: For information about the Douglas, Wyoming Jackalope Days event, held each June, see https://www.cityofdouglas.org/301/Jackalope-Days.

p. 20 A Limited, Non-Resident Jackalope Hunting License: The official Converse County, Wyoming jackalope hunting license may be found at https://www.cityofdouglas.org/DocumentCenter/View/104/Jackalope-hunting-license?bidId=.

CHAPTER 2: BELIEVING IS SEEING

p. 22 "As adults, anyone who likes to tell stories": Barry Sanders, "How to Tell a Story in America (Make It All True, Damned Near)," *North American Review* 284, no. 1 (1999), 42.

p. 22 I took the girls out to Pershing County: For more information on Pershing County, Nevada, see https://www.pershingcountynv.gov/.

p. 28 The story I have just fabricated: Scholarship on the American tall tale is voluminous. Several helpful sources that have informed my approach are Simon J. Bronner, *Following Tradition: Folklore in the Discourse of American Culture* (Logan: Utah State University Press, 1998); Carolyn S. Brown, *The Tall Tale in American Folklore and Literature* (Knoxville: University of Tennessee Press, 1987); Michael Dylan Foster and Jeffrey A. Tolbert, eds., *The Folkloresque: Reframing Folklore in a Popular Culture World* (Logan: Utah State University Press, 2015); and Barre Toelken, "Folklore in the American West," in *A Literary History of the American West* (Fort Worth: Texas Christian University Press, 1987), 45-53.

p. 28 "whales, when they have a mind to eat cod": Benjamin Franklin, *The Writings of Benjamin Franklin: Vol III,* "The Grand Leap of the Whale" (May 22, 1765), https://www.amazon.com/Writings-Benjamin-Franklin-Vol-Introduction/dp/1331607469.

p. 29 "From almost the beginning": Brown, *The Tall Tale in American Folklore and Literature*, 2.

p. 30 Our main aim is not to make sober distinctions: Toelken, "Folklore in the American West," 46.

p. 31 Video trailer for my book *How to Cuss in Western*: The trailer may be seen at https://vimeo.com/286064345.

p. 32 "The humorous story is told gravely": Mark Twain, "How to Tell a Story," in *How to Tell a Story and Other Essays* (New York: Oxford University Press, 1996), 4.

p. 32 a piece called "Jackalopes Really Exist": "Jackalopes Really Exist," *Altoona Mirror* [Altoona, PA] (March 2, 1950).

p. 32 *The New York Times* published an article: "Wyoming, Where Deer and Jackalope Play," *New York Times* (November 27, 1977), 26.

p. 34 In his *Field Guide to the American Jackalope*: Andy Robbins, *Field Guide to the North American Jackalope* (Ranchester, WY: Caput Mortuum Books, 2019).

p. 34 "about the size of a medium dog": Bill Alexander, *Everything You Always Wanted to Know about the Jackalope: But Didn't Think You Ought to Ask* (Evanston, WY: Alexander Ragtime, 1980).

p. 34 "Accounts of entire settlements": Curtis Dittwiley, "The Jackalope . . . Quo Vadis?" *Lost Springs Daily Bugle* (October 1924).

p. 35 Stories began to surface in the late 1880s: Christopher Flynn, "Story of the Jackalope," www.jackalopemovie.com/Legend/legendstorypg1.htm.

p. 36 "Jackalope milk is available": Martin, "Douglas Herrick, 82, Dies."

p. 37 The South Dakota permit allows hunting: The South Dakota jackalope hunting license may be found at https://shop.walldrug.com/products/jackalope-hunting-permit.

p. 37 the season is generally restricted: Robert Lesher, "Quest for the Warrior Rabbit," *Denver MENSA Members' Articles, Stories, and Photography*, 2000, https://numin8r.us/DENVERMA/more/jacklope.htm.

p. 38 "can also read minds": Bill Alexander, *Everything You Always Wanted to Know about the Jackalope*.

p. 38 "most serious Yeti trackers prefer": Bruce Larkin, "50 Facts about Jackalopes" (West Chester, PA: Wilbooks, 2013), 9.

p. 39 "Its existence, while improbable": Hermes Trismegistus "Hermester" Barrington, "Cenotaph of the Jackalope" (2002), https://www.oocities.org/nodotus/.

CHAPTER 3: THE CLASSIC HOAX

p. 40 "Hoaxes, to a greater or lesser degree": James Fredal, "The Perennial Pleasures of the Hoax," *Philosophy & Rhetoric* 47, no. 1 (2014): 73-97.

p. 40 the cover of the *Weekly World News*: Although early issues are not archived at the site, Bat Boy's current exploits may be followed at https://weeklyworldnews.com/.

p. 43 "a crow thieves; a fox cheats": Edgar Allan Poe, "Raising the Wind; Or, Diddling Considered as One of the Exact Sciences," *Saturday Courier* [Philadelphia] 13, no. 655 (14 Oct. 1843), 1, cols. 1-4, https://www.eapoe.org/works/tales/diddlnga.htm.

p. 44 "the defining feature of a hoax is the moment": Lynda Olman (FKA Walsh), *Sins against Science: The Scientific Media Hoaxes of Poe, Twain, and Others* (Albany: State University of New York Press, 2006), 33.

p. 44 "while the long con, when done right": Kevin Young, *Bunk: The Rise of Hoaxes, Humbug, Plagiarists, Phonies, Post-Facts, and Fake News* (Minneapolis: Graywolf Press, 2018), 96.

p. 44 "we collaborate with the hoax": Young, *Bunk*, 446.

p. 45 Orson Welles's fascinating 1973 film: *F is for Fake*, directed by Orson Welles (Paris: Planfilm Specialty Films, 1973).

p. 45 "all involve some kind of artful deception": Chris Fleming and John O'Carroll, "The Art of the Hoax," *Parallax* 16, no. 4 (2010), 45.

p. 46 "A hoax that deceives no one fails": Fredal, "The Perennial Pleasures of the Hoax," 78.

p. 48 Hoaxes take myriad forms: There is a vast literature on hoaxing, including specific studies of each of the major hoaxes I profile here. Among the most helpful general treatments of hoaxing are Alex Boese, *The Museum of Hoaxes* (New York: Plume Books, 2002); James W. Cook, *The Arts of Deception: Playing with Fraud in the Age of Barnum* (Cambridge: Harvard University Press, 2001); Leonard Diepeveen, *Modernist Fraud: Hoax, Parody, Deception* (New York: Oxford University Press, 2019); Ian Tattersall and Peter Névraumont, *Hoax: A History of Deception: 5,000 Years of Fakes, Forgeries, and Fallacies* (New York: Black Dog & Leventhal, 2018); and Young, *Bunk*.

p. 49 "GREAT ASTRONOMICAL DISCOVERIES": Quoted in Chris Wells, "The Real Moon Hoax That You Haven't Heard Of," *Beyond Bones* (June 19, 2019), https://blog.hmns.org/2019/06/the-real-moon-hoax-that-you-havent-heard-of/.

p. 50 Among my favorite examples of this is the alarmed response: For a hoax webpage detailing the supposed dangers of dihydrogen monoxide, see https://www.dhmo.org/facts.html.

p. 51 "I am impressed by the number of records of birds": MFA Meiklejohn, "Notes on the Hoodwink (*Dissimulatrix spuria*)," *Bird Notes: The Journal of the Royal Society for the Protection of Birds* XXIV, no. 5 (Summer 1950): 89.

p. 51 "generally recognizable by *blurred appearance*": Meiklejohn, "Notes on the Hoodwink," 90.

p. 51 "It is the brown blur that passes rapidly": Ibid., 89.

p. 51 "Among rows of bottles": Ibid., 90.

p. 51 "In many records of bird song": Ibid., 89.

p. 51 "Details little known, but undoubtedly": Ibid., 91.

p. 52 "a group of ornithologists": McGowan's quote is included in an email from Sankurie Pye to the author on July 23, 2020.

p. 52 "We still have the Hoodwink": Email correspondence from Sankurie Pye to the author on July 14, 2020.

pp. 52–53 As Jan Bondeson shows: Jan Bondeson, *The Feejee Mermaid and Other Essays in Natural and Unnatural History* (Ithaca: Cornell University Press, 2014). Also helpful is Steven Levi, "P. T. Barnum and the Feejee Mermaid," *Western Folklore* 36, no. 2, (1977): 149-54. For general studies of Barnum in the context of the hoax, see Benjamin Reiss, *The Showman and the Slave: Race, Death, and Memory in Barnum's America* (Cambridge: Harvard University Press, 2001); Robert Wilson, *Barnum: An American Life* (New York: Simon and Schuster, 2019); and Young, *Bunk*.

p. 55 "of all the Mammalia yet known": Quoted in Des Cowley and Brian Hubber, "Distinct Creation: Early European Images of Australian Animals," *The La Trobe Journal* 66 (Spring 2000), 19, http://www3.slv.vic.gov.au/latrobejournal/issue/latrobe-66/t1-g-t2.html.

p. 56 "it was impossible not to entertain": Quoted in Cowley and Hubber, 19.

p. 57 "Legends say that jackalopes conceive": Robbins, *Field Guide to the North American Jackalope*, 23.

p. 57 "I got interested in the jackalope through taxidermy": This and all subsequent Andy Robbins quotes are from a phone interview with the author on July 16, 2020.

p. 58 "authors a fake that reveals a deeper fake": Fredal, "The Perennial Pleasures of the Hoax," 79.

p. 59 "Folk culture in the Digital Age": Trevor J. Blank, *Folk Culture in the Digital Age: The Emergent Dynamics of Human Interaction* (Logan: Utah State University Press, 2012), 3–4.

p. 60 questions about his work on Internet folklore: This and all subsequent Trevor J. Blank quotes are from email correspondence with the author between July 25, 2020 and September 9, 2020.

p. 60 A website created by a man named Richard Reibel: All subsequent quotes from Reibel's website (including submissions by readers to the "True Stories" section of the site) are from The Jackalope Conspiracy, http://www.sudftw.com/.

p. 63 "Back in the midnineties I was trying to learn HTML": This and all subsequent Richard Reibel quotes are from a phone interview with the author on July 15, 2020.

CHAPTER 4: MOUNTING ENTHUSIASM

p. 66 "But when I make a good [taxidermy] mount": Christopher Buehlman, *Those Across the River* (New York: Ace Books, 2011), 55-56.

p. 67 True taxidermy has been practiced for centuries: For more on the history of taxidermy, see Stephen T. Asma, "Flesh-eating Beetles and the Secret Art of Taxidermy," in *Stuffed Animals and Pickled Heads: The Culture and Evolution of Natural History Museums* (New York: Oxford University Press), 3-46.

p. 68 But a third approach to animal taxidermy: For more on rogue taxidermy, see Robert Marbury, *Taxidermy Art: A Rogue's Guide to the Work, the Culture, and How To Do it Yourself* (New York: Artisan, 2014) and Kat Su, *Crap Taxidermy* (Berkeley, CA: Ten Speed Press, 2014).

p. 69 "It's really to commemorate the animal": "Brewer Defines the Art of Rogue Taxidermy," Minneapolis College of Art and Design, *The Current*, https://mcad.edu/news/brewer-defines-art-rogue-taxidermy.

p. 69 The title of an article in the *Guardian*: Kim Kelly, "'Rogue Taxidermy': A Misunderstood Ethical Art Form or the Next Hipster Fad?," *Guardian* (November 16, 2016), https://www.theguardian.com/culture/2016/nov/16/taxidermy-hipster-art-ethics-morbid-anatomy-museum.

p. 72 "Then, thirty-five years ago, I got the call": This and all subsequent Frank and Dianne English quotes are from an interview conducted with the author at their home in Rapid City, South Dakota, on October 17, 2018.

p. 79 If this massive operation is built out: Ian Palmer, "Feds Approve Plan to Drill and Frack 5,000 New Oil Wells in the Powder River Basin of Wyoming," *Forbes* (January 14, 2021), https://www.forbes.com/sites/ianpalmer/2021/01/14/feds-approve-plan-to-drill-and-frack-5000-new-oil-wells-in-the-powder-river-basin-of-wyoming/?sh=6439b9eb6bb6.

p. 80 "Come on back": This and all subsequent Michael Herrick quotes are from an interview with the author at Herrick's taxidermy shop, Antler Taxidermy and Arts, in Mills, Wyoming, on October 17, 2018.

CHAPTER 5: A MIGHTY RIVER OF JACKALOPIANA

p. 86 "So when you're searching for the truth": Okee Dokee Brothers, "Jackalope," track 5 on *Saddle Up*, Okee Dokee Music, 2016, compact disc.

p. 86 The jackalope mount is the headwaters: Two excellent sources for jackalopiana are the Wyoming gift shop Jackalope Junction (https://www.jackalopejunction.com/) and the South Dakota emporium Wall Drug (https://www.walldrug.com/).

p. 87 "kitsch seems to be in possession": Ruth Holliday and Tracey Potts, *Kitsch! Cultural Politics and Taste* (Manchester, UK: Manchester University Press, 2012), 36.

p. 87 "No matter how we scorn it": Milan Kundera, *The Unbearable Lightness of Being* (New York: Perennial Classics, 1999), 256.

p. 87 "We collect things made in earnest": Quoted in Bella English, "Doing a Good Deed with Bad Art," *Boston Globe* (February 8, 2009), http://archive.boston.com/ae/theater_arts/articles/2009/02/08/doing_a_good_deed_with_bad_art/.

p. 88 "Camp is a vision of the world": Susan Sontag, *Against Interpretation: And Other Essays* (New York: Octagon Books, 1978), 279.

p. 88 these tacky souvenirs are jackalope postcards: Two helpful sources of information on tall-tale postcards are Roger L. Welsch, *Tall-Tale Postcards: A Pictorial History* (New York: A.S. Barnes and Company, 1976) and Cynthia Elyce Rubin and Morgan Williams, *Larger than Life: The American Tall-Tale Postcard 1905–1915* (New York: Abbeville Press, 1990). The image of the rabbit being lifted with block and tackle appears in Rubin and Williams, 40.

p. 89 made available online images of almost 100: Sean O'Brien's Jackalope Postcard Museum is at http://jackalopepostcard.blogspot.com/.

p. 90 "I never set out to have the world's largest collection": This and all subsequent Sean O'Brien quotes from are from a phone interview with the author on August 24, 2020.

p. 93 "Oh, around forty years, I guess": This and all subsequent Rick and Sarah Hustead quotes are from an interview with the author at Wall Drug Store, in Wall, South Dakota, on October 16, 2018.

p. 95 "Jim Herrick delivers four hundred jackalopes": The likely Hustead-Herrick connection was verified in email correspondence with Sarah Hustead on July 12, 2021. The quote from the Herrick obituary may be found in Martin, "Douglas Herrick, 82, Dies."

p. 98 "where spontaneous order happens": Jackalope Freedom Festival Facebook page, "About," https://www.facebook.com/JackalopeFreedomFestival/.

p.99 "You fuckin' suck!": Foo Fighters, *Super Saturday Night*, YouTube video, 5:31 (January 18, 2021), https://www.youtube.com/watch?v=j8LeYu2uybU.

p. 99 "In Search of Jackalopes" tour: "Who Stole the Jackalope?" https://www.davidbowie.com/2002/2002/11/09/who-stole-the-jackalope.

p. 102 "characters or animals with crazy attributes": This and all subsequent Joe Mailander quotes are from a phone interview with the author on August 27, 2020.

p. 105 "Rabbits and hares are especially appealing": This and all subsequent Kari Serrao quotes are from email correspondence with the author between August 15 and 25, 2020.

p. 107 "the jackalopes are incredibly competent lovers": Hannah Wilde, *Violated by Monsters: The Jackalope Farm* (2014), 276, Kindle.

p. 108 "online world where you'll become": https://lutris.net/games/feral/.

p. 108 "Texas's answer to the minotaur": The Frasier Archives, Transcripts, season 7, episode 7, "A Tsar is Born," http://www.kacl780.net/frasier/transcripts/season_7/episode_7/a_tsar_is_born.html.

p. 109 "this sage of the sage": The quotes in this paragraph are from *Boundin'*, written by Bud Luckey and directed by Bud Luckey and Roger Gould (Burbank, CA: Buena Vista Pictures, 2003).

p. 110 "the lowest grossing film of all time": Plot summary, *Return of the Jackalope*, directed by Michael D. Friedman and Dave R. Watkins (Cougars Film Group, 2006), *Internet Movie Database*, https://www.imdb.com/title/tt0436717/plotsummary.

p. 110 "I'd rather do the creative part": This and all subsequent Dave R. Watkins quotes are from a phone interview with the author on September 3, 2020.

p. 111 "I love Disney movies": The quotes in this paragraph are from *Return of the Jackalope.*

CHAPTER 6: NECESSARY MONSTERS

p. 113 "It can be difficult to determine": Malcolm South, *Mythical and Fabulous Creatures: A Source Book and Research Guide* (Westport, CT: Greenwood Press, 1987), xix.

p. 114 "The Pelican of everyday zoology": Jorge Luis Borges, *The Book of Imaginary Beings*, translated by Norman Thomas di Giovanni (Boston: Dutton, 1970), 180.

p. 114 "We are as ignorant of the meaning": Borges, *The Book of Imaginary Beings*, 16–17.

p. 115 "researchers at the Biodiversity Unit": "Researchers from University of Turku Have Described Over 40 New Species in 2020," *ScienceMag* (July 1, 2020), https://scienmag.com/researchers-from-university-of-turku-have-described-over-40-new-species-in-2020/.

p. 115 Animals discovered during the past century: Among the most helpful sources on cryptozoology are Loren Coleman and Jerome Clark, *Cryptozoology A to Z: The Encyclopedia of Loch Monsters, Sasquatch, Chupacabras, and Other Authentic Mysteries of Nature* (New York: Fireside, 1999) and Darren Naish, *Hunting Monsters: Cryptozoology and the Reality behind the Myths* (London: Arcturus, 2016). Among the most comprehensive sources of information on species discovered or rediscovered in the past century is George M. Eberhart, *Mysterious Creatures: A Guide to Cryptozoology* (Santa Barbara, CA: ABC-CLIO, 2002).

p. 117 "Linnaeus even left open": W. Scott Poole, *Monsters in America: Our Historical Obsession with the Hideous and the Haunting*, 2nd ed. (Waco, TX: Baylor University Press, 2018), 11.

p. 118 the "lords of in-between": Lewis Hyde, *Trickster Makes this World: Mischief, Myth, and Art* (New York: Farrar, Straus and Giroux, 1998), 7.

p. 118 Artist Lily Seika Jones's: Jones's illustration appears as the facing page image to the jackalope chapter of Veronica Wigberht-Blackwater and Melissa Brinks, *The Compendium of Magical Beasts: An Anatomical Study of Cryptozoology's Most Elusive Beings* (Philadelphia: Running Press Adult, 2018).

p. 118 "Sometimes a writer's aim": South, *Mythical and Fabulous Creatures*, xxvi.

p. 119 "scientists still don't know how deeply": "The Museum's Giant Squid," American Museum of Natural History (24 Jan. 2013), https://www.amnh.org/explore/news-blogs/from-the-collections-posts/the-museum-s-giant-squid.

p. 119 explores deception, mischief, and misinformation: The Museum of Hoaxes, "About," http://hoaxes.org/.

p. 120 "the upland trout": Alex Boese, *The Museum of Hoaxes: A History of Outrageous Pranks and Deceptions* (New York: Plume, 2002), 81.

p. 120 "We're based in San Diego": The Museum of Hoaxes, "Finding Us," http://hoaxes.org/.

p. 120 "I still hear from school teachers": This and all subsequent Alex Boese quotes are from a phone interview with the author on July 20, 2020.

p. 122 "In 2006, a trust was set up": Merrylin Cryptid Museum, "About," http://www.merrylinmuseum.com/welcome. Subsequent quotes from the Merrylin Cryptid Museum are also from this site.

p. 123 "I definitely had an interest in unknown things": Joe Yanick, "Interview: Alex CF of The Merrylin Cryptid Collection," *Diabolique Magazine*

(April 4, 2014), https://diaboliquemagazine.com/interview-alex-cf-merrylin-cryptid-collection/.

p. 127 "The earliest-known contemporaneous report": Peter Jensen Brown, "Civic Pride through Taxidermy: A Many-Pronged History of Jackalopes," *Early Sports and Pop Culture History Blog* (September 29, 2019), https://esnpc.blogspot.com/2019/09/civic-pride-through-taxidermy-many.html.

p. 128 I located a brief article from April 16, 1933: "Horned Rabbit Exhibited," *Washington Post (1923-1954)*, 1933, R3.

p. 128 This one with a Dallas byline: Dallas Dispatch *New York Sun*, "Horned Rabbits in Texas: At Least That's the Testimony of Those Who Saw Ater's Game Bag," *Washington Post (1877-1922)*, 1915, 6.

p. 128 "Many there were who doubted the genuineness": John O'Sullivan, "Enter the War Bunny: He Wanted So Much to be Born a Billy Goat that Nature Gave Him Horns," *Forest and Stream* (April 1917), 162.

p. 129 In the June 1917 issue: "Where 'War Bunnies' Thrive: In the Lone Star State, Horned Rabbits are so Common Nobody Takes Their Picture," *Forest and Stream* (June 1917), 261.

p. 129 In a 1901 paper: Erwin H. Barbour, "A Peculiar Disease of Birds' Feet Observed in Central Nebraska," *Proceedings of the Nebraska Ornithologists' Union* 2 (1901): 61-63.

p. 129 The anomalous bunnies were also mentioned: E. W. Nelson, "The Rabbits of North America," *North American Fauna* 29 (1909), 671.

p. 129 prominent literary naturalist: Ernest Thompson Seton, *Life-Histories of Northern Animals: An Account of the Mammals of Manitoba*, vol. 1 (New York: Charles Scribner's Sons, 1909), 650-51.

p. 129 "Rabbits with horns are frequently": Seton, *Life-Histories of Northern Animals*, 672.

p. 129 Seton returned to this topic: Ernest Thompson Seton, *Lives of Game Animals*, vol. 4, pt. 2 (New York: Doubleday, Doran & Company, 1929), 811.

p. 131 which carried a Douglas byline: "Jackalopes Exist, Wyoming Claims," *Paterson Evening News* [New Jersey], (March 1, 1950).

p. 132 "2. *Undescribed, unusual, or outsize*": Eberhart, *Mysterious Creatures*, xxiii-xxiv.

p. 132 "8. *Mythical animals with a zoological basis*": Ibid.

p. 132 "10. *Known hoaxes of probable misidentifications*": Ibid.

p. 132 "The psychological significance of cryptozoology": Peter Dendle, "Cryptozoology in the Medieval and Modern Worlds," *Folklore* 117, no. 2 (2006), 190, 192.

p. 133 "The search for hidden animals is a skirmish": Quoted in Robert Michael Pyle, *Where Bigfoot Walks: Crossing the Dark Divide* (New York: Houghton Mifflin Company, 1995), 318.

p. 133 "If we manage to hang on to": Pyle, *Where Bigfoot Walks*, 17.

p. 134 International Cryptozoology Museum: https://cryptozoologymuseum.com/.

p. 134 "We include hoaxes and fakes to instill": This and all subsequent Loren Coleman quotes are from a phone interview with the author on October 23, 2020.

CHAPTER 7: THE GLOBAL JACKALOPE

p. 136 "Why pick up rabbit horns in the desert": Ken McLeod, *An Arrow to the Heart: A Commentary on the Heart Sutra* (Bloomington, IN: Trafford Publishing, 2007), 134.

p. 139 "In some lands, Hare is the messenger": Terri Windling, "The Symbolism of Rabbits and Hares" (April 24, 2012), https://www.webcitation .org/67AkEBAkB?url=http://www.endicott-studio.com/rdrm/rr Rabbits.html.

p. 141 "a yellow beast that looks like a rabbit": Quoted in Michel Wiedemann, "Les lièvres cornus, une famille d'animaux fantastiques," *Symposium* (28 March 2009), http://symposium.over-blog.fr/article-29574343 .html. Wiedemann's superb article, which is available only in French, has informed the analysis in much of this chapter. Another helpful article, available only in German, is Erwin Pokorny, "Der gehörnte Hase: Von der kaiserlichen Rarität zum Wolpertinger," in *Herrlich Wild: Höfische Jagd in Tirol* (Innsbruck, Austria: Wilfried Seipel, 2004), 64-79.

p. 141 "a divine culture hero": Stacy B. Schaefer and Peter T. Furst, *People of the Peyote: Huichol Indian History, Religion & Survival* (Albuquerque: University of New Mexico Press, 1996), 15.

p. 142 "a time when rabbit was a deer": Quoted in Jill L. Furst, "The Horned Rabbit: Natural History and Myth in West Mexico," *Journal of Latin American Lore* 15, no. 1 (1989), 143.

p. 142 "They are too heavy for me": Quoted in Furst, "The Horned Rabbit," 145.

p. 143 "You must be very clever": "The Hyena and the Hare Crash a Party," African Story Project, https://africanstoryproject.weebly.com/the-hyena -and-the-hare-crash-a-party.html.

p. 143 "Then he went to a beehive": Samuel Mabika, "Ňwampfundla," in *African Tales*, ed. Harold Scheub (Madison: University of Wisconsin Press, 2005), 168.

p. 143 "pleased to see this new kind": Mabika, "Ňwampfundla," 169.

p. 143 In a Zimbabwean variant: See Daniel S. Simberloff, "A Funny Thing Happened on the Way to the Taxidermist," *Natural History* 8, no. 87 (1987), 52.

p. 144 Ethnographer James Mooney collected: James Mooney, *Myths of the Cherokee* (Washington, D.C.: Government Printing Office, 1902).

p. 144 "When the rabbit came out": Mooney, *Myths of the Cherokee*, 276.

p. 144 "manifestations of Manabozho": For more on Manabozho, see Paul Radin and A. B. Reagan, "Ojibwa Myths and Tales. The Manabozho Cycle," *The Journal of American Folklore* 41, no. 159 (1928), 61-146.

p. 145 I received the following message: Email correspondence with the author, March 14, 2019.

p. 149 "It is noticeable among the foreign visitors": This and all subsequent Anne Blaich quotes come from email correspondence with the author between December 1 and 10, 2020.

p. 151 One of the last important manuscript illustrators: For more on Hoefnagel, see Marisa Bass, *Insect Artifice: Nature and Art in the Dutch Revolt* (Princeton, NJ: Princeton University Press, 2019).

p. 152 "those used to be little boxes": François Rabelais, *Gargantua and Pantagruel*, translated by Burton Raffel (New York: W.W. Norton & Company, 1990), 7.

p. 153 "Even as natural science matured": For helpful information on the writers mentioned in this section of the chapter, see Wiedemann, "Les lièvres cornus."

p. 153 "It is claimed that there are in Norway": Quoted in Wiedemann, "Les lièvres cornus."

p. 153 "Under the genus *Lepus* he included "Der Gehörnte Hase": Johann Christoph Heppe, *Systematisches Lehrbuch über die drey Reiche der Natur*, vol. 1 (1778), https://tinyurl.com/yu3p9mb3.

p. 155 "I am hungry and must eat": This and subsequent quotes from this tale are from Barbara O'Brien, "The Jātaka Tale of the Selfless Hare: Why There Is a Hare in the Moon," *Learn Religions* (May 8, 2017), https://www.learnreligions.com/the-jataka-tale-of-the-selfless-hare-450049.

p. 156 "One who wishes to teach others": Quoted in Hsing Yun, *The Rabbit's Horn: A Commentary on the Platform Sutra* (Hacienda Heights, CA: Buddha's Light Publishing, 2010), 43.

p. 156 "exists nowhere and clings to nothing": *The Surangama Sutra* (Len Yen Ching), Chinese rendering by Master Paramiti of Central North India at Chih Chih Monastery, Canton, China, C.E. 705, commentary (abridged) by Ch'an Master Han Shan (1546-1623), translated by Upasaka Lu K'uan Yu (Charles Luk) (London: Rider, 1966), 9.

pp. 156–57 "You just said that the nature of the knowing": *The Surangama Sutra*, 10.

p. 157 "there are some philosophers": *The Lankavatara Sutra: A Mahayana Text*, translated by Daisetz Teitaro Suzuki (London: Routledge, 1932), 46.

p. 157 "fallen into the dualistic": *The Lankavatara Sutra*, 46-47.

p. 157 "the hare's horns neither are": Ibid., 47.

p. 157 "It is the same with the hare's horns": Ibid., 48.

p. 158 "So, Sean, here's where I'm stuck": This and all subsequent quotes from the author and Sean O'Grady are from a video interview on December 18, 2020.

p. 160 "*Rabbits and horses have horns*": *The Blue Cliff Record*, vol. 2, translated by Thomas Cleary and J. C. Cleary (Boston: Shambhala, 1977), 369.

p. 161 "Some mistakenly say": *The Blue Cliff Record*, 370.

CHAPTER 8: DR. SHOPE'S WARTY RABBITS

p. 162 "The horned Cottontail is a well-known freak": Seton, *Lives of Game Animals*, 787.

p. 162 "killed more people in a year": John M. Barry, *The Great Influenza: The Story of the Deadliest Pandemic in History* (New York: Viking, 2004), 5. See also Lara Spinney, *Pale Rider: The Spanish Flu of 1918 and How it Changed the World* (New York: PublicAffairs, 2017), 37.

p. 162 "Their skin turned dark blue": Catherine Arnold, *Pandemic 1918: Eyewitness Accounts from the Greatest Medical Holocaust in Modern History* (New York: St. Martin's Press, 2018), 10-11.

p. 163 "The disease proved so severe": Douglas Jordan, "The Deadliest Flu: The Complete Story of the Discovery and Reconstruction of the 1918 Pandemic Virus," *Centers for Disease Control and Prevention*, https://www.cdc.gov/flu/pandemic-resources/reconstruction-1918-virus.html.

p. 163 The prevailing theory: Arnold, *Pandemic 1918*, 276-77.

p. 163 Climate may also have played a role: Erin Blakemore, "Catastrophic Effect of 1918 Flu May Have Been Aided by Peculiar Influx of Cold Air into Europe during WWI," *Washington Post* (October 3, 2020), https://www.washingtonpost.com/science/spanish-flu-1918-climate/2020/10/02/9b730432-0339-11eb-a2db-417cddf4816a_story.html.

p. 164 Born on Christmas Day: The most helpful biographical sources on Richard E. Shope are the following: Christopher Andrewes, "Richard Edwin Shope 1901-1966," *National Academy of Sciences* (1972), 352-75; Russell W. Currier, "Iowa's Richard Edwin Shope MD: His Contributions to Influenza Research and One Medicine/Health" (January 26, 2017); Russell W. Currier, "Richard Edwin Shope, MD, 1901-1966: Highlights of a Life of Accomplishment in the Field of Animal Health," *Veterinary Heritage: Bulletin of the American Veterinary History Society* 40, no. 2 (2017), 37-43; "Dr. Richard Edwin Shope Dead; Pathologist at Rockefeller U., 64," *New York Times* (October 3, 1966), 47; C. M. MacLeod, "Richard Edwin Shope: 1901-1966," *Transactions of the Association of American Physicians* 84 (1971): 36-38; "Obituary for Richard E. Shope (Aged 64)," *Des Moines Register* (October 4, 1966), https://desmoinesregister.newspapers.com/clip/59025152/obituary-for-richard-e-shope-aged-64/; Peyton Rous, "Presentation of the Kober Medal to Richard Shope," *Transactions of the Association of American Physicians* 70 (1957), 29; Nancy Shope, "A Biographical Sketch of Richard E. Shope, MD," *Veterinary Heritage: Bulletin of the American History Society* 40, no. 2 (November 2017), 66-71; Ton van Helvoort, "Shope, Richard Edwin (25 December 1901-02 October 1966)," *American National Biography* (February 2000), https://www.anb.org/view/10.1093/anb/9780198606697.001.0001/anb-9780198606697-e-1201635;jsessionid=FB53D2B08FEA2360A450656B.C.E.DBF765; and, Greer Williams, "Shope and His Shoats," in *Virus Hunters* (New York: Knopf, 1959), 200-09.

p. 165 "We certainly are getting the laughs": Here and throughout this chapter, quotes from letters written by Richard Shope are from unpublished correspondence, shared with the author by Shope's daughter, Nancy Helen Shope FitzGerrell.

p. 167 "A careful investigation would seem warranted": Richard E. Shope, "Swine Influenza III: Filtration Experiments and Etiology," *Journal of Experimental Medicine* 54, no. 3 (1931), 384.

p. 167 In his subsequent research: For Shope's main article reporting the findings discussed in this paragraph, see Richard E. Shope, "The Incidence of Neutralizing Antibodies for Swine Influenza Virus in the Sera of Human Beings of Different Ages," *Journal of Experimental Medicine* 63, no. 5 (1936), 669-84.

p. 168 Having returned to the Rockefeller Institute: Nancy Shope, "A Biographical Sketch," 69.

p. 168 "I have had a bunch": Unpublished letter, January 18, 1932.

p. 169 enormous "human importance": Rous, "Presentation of the Kober Medal," 32.

p. 169 In 1952 French bacteriologist: "Paul-Félix Armand-Delille," Wikipedia, https://en.wikipedia.org/wiki/Paul-F%C3%A9lix_Armand-Delille.

p. 169 "referred to popularly": Shope, "Infectious Papillomatosis of Rabbits," 607.

p. 170 "The father of the wife": Quoted in Ludwik Gross, *Oncogenic Viruses* (Oxford, UK: Pergamon Press, 1961), 27.

p. 170 "We had a young boy": Quoted in Gross, *Oncogenic Viruses*, 27-28.

p. 171 Once rabbits were infected: Shope, "Infectious Papillomatosis of Rabbits," 617. See also Richard E. Shope, "Immunization of Rabbits to Infectious Papillomatosis," *Journal of Experimental Medicine* 65 (1937), 219-31.

p. 171 "Rabbits carrying experimentally": Ibid., 623.

p. 173 "the greatest joys and rewards": MacLeod, "Richard Edwin Shope," 36.

p. 173 "gained a new dimension": Andrewes, "Richard Edwin Shope," 361.

p. 173 "the fourth new disease": Rous, "Presentation of the Kober Medal," 32.

p. 173 "Rous's note simply reads": Ibid., 33.

p. 174 "The work that Rous and others": Richard E. Shope, "Evolutionary Episodes in the Concept of Viral Oncogenesis," *Perspectives in Biological Medicine* 9, no. 2 (Winter 1966), 265.

p. 174 "He used to call me his Little Shadow": This and all subsequent Thomas Shope quotes are from a phone interview with the author on April 12, 2021.

p. 176 "My dad was very entertaining": This and all subsequent Nancy Shope FitzGerrell quotes are from a phone interview with the author on April 12, 2021.

p. 178 "I have chosen R. E. Shope": Nancy Shope, "A Biographical Sketch of Richard E. Shope," 71.

p. 180 "The jackalope may be an example": Simberloff, "A Funny Thing Happened on the Way to the Taxidermist," 54.

CHAPTER 9: SAVED BY JACKALOPES

p. 181 "Tumors destroy man": Peyton Rous, "The Challenge to Man of the Neoplastic Cell (Peyton Rous Nobel Lecture)," *The Nobel Prize* (December 13, 1966), https://www.nobelprize.org/prizes/medicine/1966/rous/lecture/.

p. 181 The year after Rous's arrival: H. Cody Meissner, "From Peyton Rous to the HPV Vaccine: A Journey of Discovery and Progress," *Pediatrics: Official Journal of the American Academy of Pediatrics* 144, no. 6 (December 1, 2019), https://pediatrics.aappublications.org/content/144/6/e20192345.

p. 182 "Most scientists viewed cancer cells": "1966 Nobel Prize in Physiology or Medicine," *The Rockefeller University*, https://www.rockefeller.edu/our-scientists/peyton-rous/2493-nobel-prize/.

p. 182 "cancer researchers and physicians": Neeraja Sankaran and Ton van Helvoort, "Andrewes's Christmas Fairy Tale: Atypical Thinking about Cancer Aetiology in 1935," *Notes and Records of the Royal Society of London* 70, no. 2 (March 2016), 3.

p. 183 "Shope remarked that they might be": Rous, "The Challenge to Man of the Neoplastic Cell," 27.

p. 183 "Dr. Rous persisted in this line": "1966 Nobel Prize in Physiology or Medicine."

p. 183 "HURRAY! They should have done it": Quoted in Sankaran and van Helvoort, "Andrewes's Christmas Fairy Tale," 17.

p. 184 It was later named: Nancy M. Cladel, et al., "The Rabbit Papilloma Virus Model: A Valuable Tool to Study Viral-Host Interactions," *Biological Sciences* 374, no. 1773 (May 27, 2019), 2.

p. 184 Highly valued for its reproducibility: Cladel, et al., "The Rabbit Papilloma Virus Model," 7.

p. 184 "Animal models are essential": Janet L. Brandsma, "The Cottontail Rabbit Papillomavirus Model of High-Risk HPV-Induced Disease," *Methods in Molecular Medicine* 119 (2005), 217.

p. 184 Isabelle Giri and her colleagues: Isabelle Giri, et al., "Genomic Structure of the Cottontail Rabbit (Shope) Papillomavirus," *Proceeding of the Natural Academy of Sciences of the United States of America* 82, no. 6 (1985), 1580-84.

p. 184 "Once in a while I'd come across": This and all subsequent Robert Timm quotes are from email correspondence with the author on July 13, 2020.

p. 185 "working with a team of molecular biologists": Clara Escudero Duch, et al., "A Century of Shope Papillomavirus in Museum Rabbit Specimens," *PLoS One* 10, no. 7 (2015), e0132172.

p. 186 (rarely) fish: T. Mizutani, "Papillomaviruses and Polyomaviruses in Fish," *Aquaculture Virology* (2016), 177-81.

p. 186 sprouting from the back of his head: For more on the examples of human horns discussed in this paragraph, see Diane Mapes, "These Aren't Devil's Horns: They're Real!" *NBC: The Body Odd* (15 April 2009), http://www.medirabbit.com/EN/Skin_diseases/Viral_diseases/Pap/Papilloma.htm.

p. 186 the horrific story of Dede: Carl Zimmer, *A Planet of Viruses*, 3rd ed. (Chicago: University of Chicago Press, 2021), 34.

p. 187 Even dinosaurs had warts: Roger Highfield, "Is This the Oldest Human Virus? The Papillomavirus Has Been Afflicting Humans and their Ancestors for Millions of Years," *Daily Telegraph* (July 18, 2006), https://www.telegraph.co.uk/technology/3346456/Is-this-the-oldest-human-virus.html.

p. 187 "the established nucleotide variations": Ariana Harari, Zigui Chen, and Robert D. Burk, "Human Papillomavirus Genomics: Past, Present and Future," *Current Problems in Dermatology* 45 (2014), 1.

p. 187 modern humans probably received some HPV strains: Ville N. Pimenoff, et al., "Transmission between Archaic and Modern Human Ancestors during the Evolution of the Oncogenic Human Papillomavirus 16," *Molecular Biology and Evolution* 34, no. 1 (January 2017), 4-19. Also see Zigui Chen, et al., "Niche Adaptation and Viral Transmission of Human Papillomaviruses from Archaic Hominins to Modern Humans," *PLoS Pathogens* 14, no. 11 (2018), e1007352.

p. 187 "Yes, Humans and Neanderthals": Brian Resnick, "Yes, Humans and Neanderthals Had Sex. And They Gave Us an STD," *Vox* 21 (October 2016), https://www.vox.com/science-and-health/2016/10/19/13324842/neanderthals-hpv-genital-warts.

p. 187 Compare HPV's extraordinary pedigree: Highfield, "Is This the Oldest Human Virus?"

p. 188 According to the CDC's analysis for 2018: "Sexually Transmitted Infections Prevalence, Incidence, and Cost Estimates in the United States." CDC website, https://www.cdc.gov/std/statistics/prevalence-incidence-cost-2020.htm.

pp. 188–89 The CDC expresses this prevalence: "Human Papillomavirus (HPV) Statistics," CDC website, https://www.cdc.gov/std/hpv/stats.htm.

p. 189 The good news is that: "Human Papillomavirus (HPV) and Cervical Cancer," WHO (November 11, 2020), https://www.who.int/news-room/fact-sheets/detail/human-papillomavirus-(hpv)-and-cervical-cancer.

p. 189 Unfortunately, some HPV infections: For more on the statistics offered in this paragraph, see "Papillomaviruses and Human Cancer," *ViroBlogy* (November 3, 2015), https://rybicki.blog/2015/02/11/papillomaviruses-and-human-cancer/.

p. 189 Despite the resistance: Zimmer, *Planet of Viruses*, 41.

p. 189 GLOBOCAN cancer statistics: Hyuna Sung, et al., "Global Cancer Statistics 2020: GLOBOCAN Estimates of Incidence and Mortality Worldwide for 36 Cancers in 185 Countries," *CA: A Cancer Journal for Clinicians* 71, no. 3 (2021), 209-49, doi:10.3322/caac.21660.

p. 190 "The global disparity in cervical cancer": Christopher P. Wild, Elisabete Weiderpass, and Bernard W. Stewart, eds., *World Cancer Report: Cancer Research for Cancer Prevention* (2020), 394, https://publications.iarc.fr/586.

p. 190 "Inequities and disparities": "Cervical Cancer: Statistics," Cancer.Net (2021), https://www.cancer.net/cancer-types/cervical-cancer/statistics.

p. 190 The cause of cervical cancer: For more on the statistics offered in this paragraph, see Jennifer Houtman, "Viruses, Cancer, Warts and All: The HPV Vaccine for Cervical Cancer," *Federation of American Societies for Experimental Biology* (5 June 2008), 12, https://www.faseb.org/portals/2/pdfs/opa/2008/HPV.pdf.

p. 191 "My hypothesis was based on several": Harald zur Hausen, "HPV Vaccines: What Remains to be Done?" *Expert Reviews* (2011), 1505.

p. 191 "indicated that there would be no market": Harald zur Hausen, *Nobel Lectures in Physiology or Medicine (2006–2010)*, ed. Göran K. Hansson (Singapore: World Scientific, 2014), 174.

p. 191 Despite the initial lack of enthusiasm: For more on the HPV vaccine development race, see Norma Erickson, "Behind the Scenes: The Evolution of an HPV Industry," SaneVax, http://sanevax.org/behind-the-scenes-the-evolution-of-an-hpv-industry/. Also see Brad Rodu, "Inventors of the Human Papillomavirus Vaccine," *R Street* (December 20, 2017), https://www.rstreet.org/2017/12/20/inventors-of-the-human-papillomavirus-vaccine/.

pp. 192–93 sales of Gardasil alone topped $3 billion: Team Trefis, "Merck's $3 Billion Drug Jumped To 4x Growth Over Previous Year," *Forbes* (October 4, 2019), https://www.forbes.com/sites/greatspeculations/2019/10/04/mercks-3-billion-drug-jumped-to-4x-growth-over-previous-year/?sh=7ac48eda6294.

p. 193 cut cervical cancer rates nearly in half: Francesca Tomasi, "An HPV Feature: From Jackalope to Cancer Vaccines," *Infective Perspective* (2016), http://www.infectivepersective.com/blog/an-hpv-feature-from-jackalope-to-canver-vaccine.

p. 193 "Eliminating any cancer": Quoted in Manjulika Das, "WHO Launches Strategy to Accelerate Elimination of Cervical Cancer," *The Lancet Oncology* 22, no. 1 (2021) 20-21, doi:10.1016/S1470-2045(20)30729-4.

p. 193 In the fifteen years since its introduction: For more information on the HPV vaccine roll out, see Hallie Whitman and Stephane Cajigal, "Timeline: 10 Years of the HPV Vaccine," *Medscape* 5 (August 2016), https://www.medscape.com/viewarticle/866964.

p. 194 "Both the estimated uptake": Thinley Dorji, et al., "Human Papillomavirus Vaccination Uptake in Low-and Middle-Income Countries: A Meta-Analysis," *EClinicalMedicine* 34 (April 1, 2021), https://www.sciencedirect.com/science/article/pii/S2589537021001164.

p. 195 "most often refused of common vaccines": Whitman and Cajigal, "Timeline: 10 Years of the HPV Vaccine."

p. 195 "Here in the United States": This and all subsequent Elisa Sobo quotes are from a phone interview with the author on June 9, 2021.

p. 197 "HPV-caused cancers are a little like tobacco": This and all subsequent John Hess quotes are from an in-person interview with the author in Reno, Nevada on June 10, 2021.

p. 199 "published her journal as the book": Julie Forward DeMay, *Cell War Notebooks: My Journey with Cervical Cancer* (CreateSpace Independent Publishing Platform, 2011).

p. 200 "Being bald is the cancer signal": DeMay, *Cell War Notebooks*, 23.

p. 200 "I continue to write in my journal": Ibid., 73.

p. 200 "Today I walked down to work": Ibid., 15.

p. 200 "Cervical cancer research has come a long way": Ibid., 41.

p. 201 "She was a very free spirit": This and all subsequent Jane Forward quotes are from a phone interview with the author on June 9, 2021.

p. 202 "It's just time to let go": DeMay, *Cell War Notebooks*, 23.

CHAPTER 10: THE JACKALOPE MAKER

p. 203 "Stuffed animals are cute": Demetri Martin, *Live (At the Time)*, directed by Jay Karas, Netflix, 2015.

p. 204 "curiously mind-expanding treasures": https://paxtongate.com/.

p. 206 "I see roadkill all over the place": This and all subsequent quotes in this chapter are close approximations of the comments and conversations that occurred during the jackalope making workshop.

p. 209 "577,588 such deaths in 2020 alone": Sung, et al., "Global Cancer Statistics 2020." The figure of 577,588 is arrived at by adding the GLOBOCAN's projected deaths for the following HPV-caused cancers: cancers of the vagina, vulva, anus, and penis (58,000); oral cavity cancers (77,757); and cervical cancer (341,831).

p. 218 my favorite San Francisco bar: Jackalope: http://www.jackalope-sf.com/.

Index

H

I

J

Y

About the Author

Michael P. Branch is University Foundation Professor of English and Professor of Literature and Environment at the University of Nevada, Reno. He currently serves as the Warren B. Lepus Chair of Field Research at the International Institute for Jackalope Studies, headquartered in Geneva. His nine books include three works of humorous creative nonfiction inspired by the Great Basin Desert: *Raising Wild* (2016), *Rants from the Hill* (2017), and *How to Cuss in Western* (2018). He has published more than 300 essays and reviews which have appeared in venues including *Orion*, *CNN*, *Slate*, *Outside*, *Pacific Standard*, *Utne Reader*, *National Parks*, *Ecotone*, *High Country News*, *Terrain.org*, *Places Journal*, *Bustle*, *Whole Terrain*, and *About Place*. His nonfiction includes pieces that have been recognized as Notable Essays in *The Best American Essays*, *The Best Creative Nonfiction*, *The Best American Science and Nature Writing*, and *The Best American Nonrequired Reading*. He is the recipient of the Ellen Meloy Desert Writers Award, the Nevada Writers Hall of Fame Silver Pen Award, the Western Literature Association Frederick Manfred Award for Creative Writing, and the Montana Prize for Humor. Mike lives with his wife, Eryn, and daughters, Hannah and Caroline, in the ecotone where the western Great Basin Desert and the eastern slope of the Sierra Nevada Mountains meet.